ISBN	Author	Title
0-07-042591-4	Minoli/Vi	
0-07-044362-9	Muller	
0-07-912256-6	Naugle	
0-07-046461-8	Naugle	*Network Protocol Handbook*
0-07-046380-8	Nemzow	*The Ethernet Management Guide,* 3/e
0-07-046321-2	Nemzow	*The Token-Ring Management Guide*
0-07-049663-3	Peterson	*TCP/IP Networking: A Guide to the IBM Environment*
0-07-051143-8	Ranade/Sackett	*Advanced SNA Networking: A Professional's Guide to VTAM/NCP*
0-07-051506-9	Ranade/Sackett	*Introduction to SNA Networking,* 2/e
0-07-053199-4	Robertson	*Accessing Transport Networks*
0-07-054991-5	Russell	*Signaling System #7*
0-07-054418-2	Sackett	*IBM's Token-Ring Networking Handbook*
0-07-057199-6	Saunders	*The McGraw-Hill High-Speed LANs Handbook*
0-07-057442-1	Simonds	*McGraw-Hill LAN Communications Handbook*
0-07-057639-4	Simonds	*Network Security: Data and Voice Communications*
0-07-060363-4	Spohn	*Data Network Design,* 2/e
0-07-069416-8	Summers	*ISDN Implementor's Guide*
0-07-063263-4	Taylor	*The McGraw-Hill Internetworking Handbook*
0-07-063295-2	Taylor	*Multiplatform Network Management*
0-07-063638-9	Terplan	*Benchmarking for Effective Network Management*
0-07-063639-7	Terplan	*Effective Management of Local Area Networks: Functions, Instruments and People,* 2/e
0-07-065766-1	Udupa	*Network Management System Essentials*
0-07-067375-6	Vaughn	*Client/Server System Design and Implementation*

To order or to receive additional information on these or any other McGraw-Hill titles, please call 1-800-822-8158 in the United States. In other countries, contact your local McGraw-Hill representative.

KEY = WM16XXA

SNA, APPN, HPR, and TCP/IP Integration

SNA, APPN, HPR, and TCP/IP Integration

David G. Matusow
Hypercom Network Systems

McGraw-Hill

New York San Francisco Washington, D.C. Auckland Bogotá
Caracas Lisbon London Madrid Mexico City Milan
Montreal New Delhi San Juan Singapore
Sydney Tokyo Toronto

Library of Congress Cataloging-in-Publication Data

Matusow, David G.
SNA, APPN, HPR, and TCP/IP integration / David G. Matusow.
p. cm. — (McGraw-Hill series on computer communications)
Includes index.
ISBN 0-07-041051-8 (hc : alk. paper)
1. Local area networks (Computer networks) I. Title.
II. Series.
TK5105.7.M388 1996
004.6′8—dc20 95-50879
CIP

1 2 3 4 5 6 7 8 9 0 DOC/DOC 9 0 1 0 9 8 7 6

ISBN 0-07-041051-8

The sponsoring editor for this book was Jerry Papke, the editing supervisor was Bernard Onken, and the production supervisor was Suzanne W. B. Rapcavage. It was set in Century Schoolbook by Estelita F. Green of McGraw-Hill's Professional Book Group composition unit.

Printed and bound by R. R. Donnelley & Sons Company.

This book is printed on recycled, acid-free paper containing a minimum of 50% recycled, de-inked fiber.

To Phyllis, Justin, and Chris. I couldn't have completed this work without their patience and love.

Contents

Acknowledgments

Many people assisted me in this effort during the 15 months of work. Although I have done several articles for publication, this was the first time that I have attempted to create a book.

Jay Ranade approached me to write this book. I must thank Jay for always being available for questions about the procedures necessary to complete this work.

Jerry Papke, Senior Editor at McGraw-Hill, also helped me in steering through the potential roadblocks of this effort. His assistant, Donna Namorato, was always helpful in getting me in touch with Jerry or one of the other fine people at McGraw-Hill. Joe Rivellese worked hard to get my graphics into the book.

Special thanks go to Bernie Onken, who put the book together and assisted me in overcoming any hurdles along the way. Bernie was always patient and helpful in getting the "little obstacles" out of the way so this book could be completed.

Dale Kurshner and Tracy Shiroma of the Technical Publication department at Hypercom were invaluable in making sure that my manuscript was readable. They worked long and hard and were always cheerful while helping me meet my deadlines. I have worked with both of them for many years, and their dedication and excellence was helpful.

Noor Chowdhury, Jon Young, Steve Baechle, Larry Bush, and Gary Sweeney of Hypercom reviewed many chapters of the manuscript. They were always happy to help . . . even when they didn't have the time to spare. I must also thank Paul Wallner for his review of several chapters. As the co-founder of Hypercom, current President of Hypercom Network Systems, and my manager, Paul was always busy, but found time to support me in this effort.

Marcia Peters of IBM provided excellent and extensive edits of my chapters, especially the ones concerning APPN and HPR. Anura Guruge assisted me with the work on frame relay and RFC 1490. He also provided a good sounding board for the comparison of the different protocols. I must also thank Paul Moscicki for his diagnostic work on the Legacy TCP Gateway.

David G. Matusow

Trademarks

ESCON, AIX, and RISC System/6000 are trademarks of International Business Machines Corporation.

IBM, MVS, OS/2, Application System/400, and AS/400 are trademarks of International Business Machines.

NetView and NetView/600 are trademarks of International Business Machines.

VTAM, Advanced Peer to Peer Networking, APPN, CICS, and NetView are registered trademarks of Internations Business Machines.

cisco is a trademark of cisco Systems, Inc.

Ethernet is a registered trademark of Xerox Corporation.

UNIX is a registered trademark of UNIX System Laboratories Inc.

Hypercom, Integrated Enterprise Network, and IEN are trademarks of Hypercom Incorporated.

UTS is a registered trademark of Amdahl Corporation.

Chapter

1

Subarea SNA

1.1 Definition of the Term SNA

System Network Architecture (SNA) is a network architecture that encompasses a family of associated pieces of architecture. Together, these pieces define a method of communication responsibilities that allowed development of a comprehensive network product line.

SNA had at its foundation the segmentation of the network into *domains*. Each domain was identified as a separate, self-delineated piece of the overall network. As such, each piece was a *subarea* of the whole network.[1]

1.2 History

SNA was announced in 1974. The proposed purpose of subarea SNA was to enable the International Business Machine (IBM) customer base to better support the broadening range of communication requirements. These requirements included:

- *An ever-widening range of communications.* The IBM customer base was experiencing an increased need for communication products. The customer networks were quickly expanding past the point that the existing communication methods would comfortably support.
- *An increasingly complex assortment of communication interfaces.* Customers were creating more and more complex networks that

[1]The term *subarea* has a more exact definition that will be discussed later in this chapter.

provided their businesses a competitive advantage. The advantage was seen as the ability to allow communication between an increasingly diverse customer base.[2]

The networks that needed to be supported by the customer base included such items as dumb terminals (3270), printers, and remote job entry (RJE) devices. Although software existed to enable these components, as the networks became large, the ability of the data center to support the pieces became more and more difficult.

- *Support and development costs associated with the existing product line.* IBM was finding it more costly to support the product development necessary to support the existing product base.
- *Duplication of effort.* IBM found that they were duplicating efforts in the development of communications interfaces. With the software becoming more complicated, IBM was finding that instead of having to pay for the development of an interface once, it was paying repeatedly. In addition, IBM was not gaining an advantage by building upon a solid base. Instead, because the groups that were doing development did not use a common code base, a correction for one product was not reflected in a different product. Thus, the customer was having to isolate and apply corrections against multiple products.

Communication development had been done by each of the program products. If a function was required, it was created by the development team associated with the product. When this same function was required by another product, it was started over again without regard to the first product's development.

This lead to high costs, long lead times, poor integration, and a low level of coordination. Experts were positioned throughout the company for each interface. Duplication was rampant and products were limited in their ability to support new interface types.

What was needed was a method of allowing a coordinated front to their customers and a way of reducing duplication of development. IBM also needed the ability to build upon the efforts of corporate experts in a specific technology.

The use of corporate experts was seen as enabling the development of a common direction for the company to its customers and better and faster support for those customers. Together, this becomes a more

[2]Here the customer is the IBM customer.

efficient and effective method, and helped IBM to enable more consistency and expansion capability in its network offerings.

1.3 Architecture

SNA was one of the first architected communication interfaces. Though the architecture was modified to make it more consistent with the International Organization for Standardization (ISO) Open System Interconnection (OSI) seven-layered model, the fundamental layers have been largely unchanged in the last 20 years.

The use of an architected model allowed each layer to become independent of each other. Modifications could be applied to a layer without resulting in a change to any other layer.

By designing the interfaces between layers well, an architecture provides a simple method of enabling cooperative development by allowing groups to develop each layer (or a portion of a layer) with little regard to the development of other layers. The result of this effort is to have the interfaces between the layers well enough defined to enable layers to work together once they are joined.

SNA is composed of many architectural components.[3] The main topics that will be discussed include:

- Network addressable units
- Network addressing
- Subareas
- Node and node types
- Subareas

1.3.1 Network addressable units

As is true of people, network components must be addressable to each other. The form of the address varies, as we shall see, but it is only the addressability that enables components to pass information between each other.

People use names. By using this method of logically identifying each other, we are able to direct information between us. Without this identification method, we would need to use physical interaction to serve the same purpose. For example, we might have to tap each

[3]Many of these components are quite complex. I will not provide a detailed discussion of these in this book.

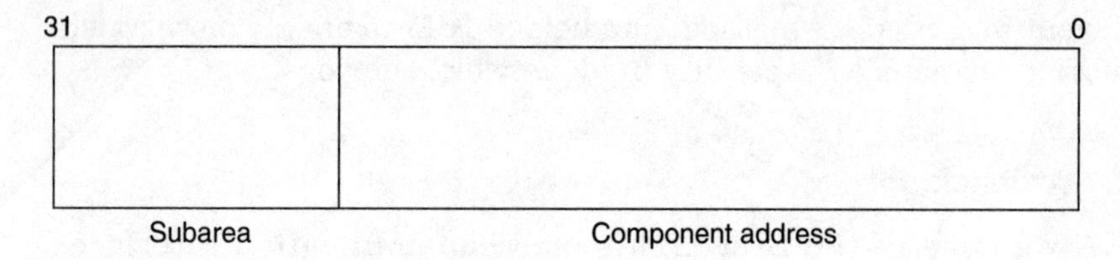

32 bit network address that consists of 2 parts. These are:

- Subarea
- Component

The *subarea* portion defines a group of components within the network. This is used to sub-group the network into pieces. The *component* portion defines the component within the subarea.

As the address is fixed in size, as the portion of the address that is used for subarea increases, the number of components within each subarea decreases.

Figure 1.1 Subarea SNA address format.

other on the shoulder before passing information. Without taking this step, our communications, such as verbal speaking, would often fall on deaf ears.

1.3.2 Network addressing

Computers can use a person-friendly interface of names, but must first translate those names into a form easier to work with for itself. In the case of SNA, this form is a unique identifier for each component within the network.

For subarea SNA, this is a 32-bit address. Each component that needs to pass information is assigned a unique address. It answers to *only* that address for communication. Figure 1.1 shows this address and how it is divided into subarea[4] and component portions.

Since all components within subarea SNA are predefined,[5] the system generation process creates a table that allows humanly usable names to be associated with a particular network component or address. The process is actually very simple.

---

[4]This is how subarea SNA gets its name.

[5]There are methods of using dynamic definitions, such as are used for cross-domain components, but I will not go into this topic in much depth in the present book.

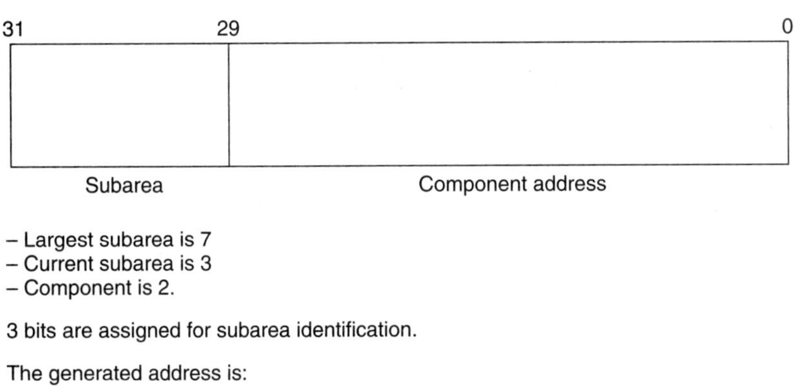

31
29
0
Subarea
Component address
– Largest subarea is 7
– Current subarea is 3
– Component is 2.
3 bits are assigned for subarea identification.
The generated address is:

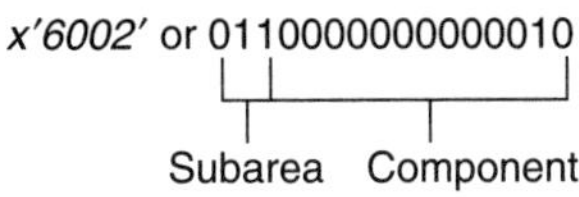

Figure 1.2 Subarea SNA address example.

Since the maximum subarea within the network is defined, the number of bits associated with the subarea portion of the network address is known. With this information and the subarea number that is being defined, the generation process simply increments a counter that allows addressing to be done. Thus, the second addressable item in the generation is given the component address of 2 within the subarea.

Figure 1.2 shows how the address for this component would be generated. Note that if more components are added before this one, the generation process must be rerun and new addresses assigned. From Figure 1.2, you can see that addressing, if a correlation table was not created and used, would not be very user-friendly.

1.3.3 Subareas

As we have seen, the network address is composed of a subarea and component address. The subarea portion of the address allows components that are related to be grouped together.

Subareas also have significance because each Virtual Telecommunications Access Method (VTAM) and Network Control Program (NCP) defines a different subarea. As a result of the fixed length of the entire address, as the network contains more subarea nodes, it is possible that the portion of the network address that must be used to define the subarea may need expansion. Such a situation can have adverse effects if each of the subareas are at capacity for components because the network address cannot be generated. This is because

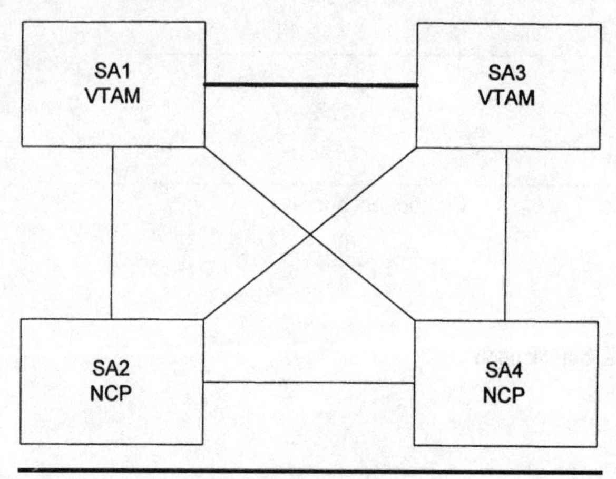

Figure 1.3 Multidomain subarea network.

more bits must be used for the subarea portion, not having sufficient bits for component addresses.[6]

Figure 1.3 shows a multidomain network.[7] This network contains four subareas (SA1, SA2, SA3, and SA4). SA1 and SA3 represent VTAM hosts. SA2 and SA4 illustrate NCPs. The links between each of these provides communication between the subareas. These cross-domain links have special significance and have characteristics that will be discussed in later chapters.

1.3.4 Nodes and node types

SNA categorizes components into different types. This architected method allows components to be understood before communication is actually established. By knowing the basic behavior of a component, it is possible to support new physical devices, as long as they adhere to the method of communication of a known node type.

The types of nodes within SNA are:

- Physical unit (PU)
- Logical unit (LU)

---

[6]Later versions of subarea SNA addressed this by expanding the size of the subarea address to use a full 32 bits and the component portion to a full 32 bits. This 64-bit address provided the expansion necessary as SNA networks grew.

[7]This can also be called a *multisubarea* network.

Among these two types, there are several subdivisions. This classification provides the ability to better understand the operation of a node.

1.3.4.1 Physical unit. The physical unit (PU) is often misunderstood. Largely because the word *physical* is in the name, it is often thought that a PU represents a physical device. This is not true!

A PU is the management entity for the LUs that are subordinant to a PU. It is the management function that is unique to a PU, not the fact of being able to hold it.

PUs are subdivided into five basic types. These are:

- PU type 5
- PU type 4
- PU type 2.0
- PU type 2.1
- PU type 1[8]

PU 5. PU type 5 (I will use PU5) are normally located in host systems. A PU5 has both routing and session management functions. The routing function allows decisions as to which path between two points in the network is to be used. The session management controls the establishment and destruction of sessions.

The session requests are passed to the PU5, which makes a determination of where the destination is located, any special considerations for the session (such as security or throughput requirements), and initiates the steps for the session to be established. The PU5 is often called system services control point (SSCP).

PU 4. PU4 has all of the functions of a PU5, except for the session management. PU4 is able to provide routing services and some independent management services. These management services are done as delegated by the SSCP.

Management is normally done between a PU and an SSCP.[9] This flow is on the SSCP-PU session.

PU 2.0. This is a peripheral PU that relies upon an SSCP for routing services. It uses a logical hierarchical network topology, such as exists within a subarea SNA environment.

---

[8]We will not be discussing PU type 1, as this is an older type that is not of interest here.

[9]Note that an SSCP is a specialized PU.

The PU 2.0 is reliant upon an SSCP for services for both itself and its subordinant LUs. The PU 2.0 node cannot itself provide most of the session services expected by its attached LUs.

PU 2.1. This is a peripheral PU that does not need to use an SSCP for routing services. PU 2.1 nodes can support *peer-peer* connections between LUs subordinate to the PU 2.1 mode. These nodes normally can operate as either primary or secondary link stations, on a dynamic basis. It is necessary to utilize this node type in order to support parallel links from a peripheral node.

The PU 2.1 node is able to provide session services to its associated LUs. Thus, when an LU requests a session with another LU, it passes the request to the PU 2.1 interface. This node then makes the determination of whether the destination is offhost and how to reach the destination.

Because the PU 2.1 is able to manage multiple link stations, it can determine which of the available paths to utilize to a destination. This expands the peripheral node into a routing node. This can greatly aid in creating flexible networks that can dynamically determine routes, thus increasing the availability of the network.

1.3.4.2 Logical units. Logical units (LU) are the resource endpoints of the network. Sessions that carry user data are established between LUs. Just as with PUs, LUs are not actually differentiated by being real (physical) or imaginary (logical). Instead, the delineation is made by the tasks that the LU performs.

Session endpoints can provide a diverse range of services. As a result, architecture designated several different types of LUs. These include, but are not limited to:

- LU0
- LU2
- LU6.1
- LU6.2

Each of these LU types provides a different type of service. In addition, within each of these LU types exist many variations. These differences are caused by differences in implementations and are exhibited through some of the parameters that are used to communicate session parameters, such as BIND commands.

The LU types provide a framework within which vendors can support different types of resources. Thus, although LUs of a type can support different types of services, subarea architecture dictates that some of the services are required for that LU type.

LU0. The LU type 0 (LU0) is a catch-all LU type that provides those services that the other LU types do not. Even so, for example, some attributes are considered to be "standard" for an LU0. These characteristic include:

- No bracketing
- No chaining support
- Full-duplex transmission for high throughput

Examples of common LU type 0 includes automatic teller machines (ATMs) and some RJE devices.

LU2. LU2 is the predominant type of LU. LUs within this type include the omnipresent 3270 and most emulation facilities. The requirements for LU2 designation include:

- Support for bracketing and change of direction indicators
- Support for the CANCEL command
- Half-duplex transmission

LU2 devices number in the millions worldwide. Because this type of LU dominates SNA networks, many vendors have created emulations of this LU type. By being emulated, non-IBM vendors are able to be a full member of an SNA network.

LU6.1. An LU6.1 is an older method of providing peer-to-peer connectivity. This type of session is characterized by the ability to start components within remote LUs. These components may include programs and other communication interfaces.

The LU6.1 was used by such subsystems as Customer Information Control System (CICS) to support communication between different CICS systems. By using the checkpointing of an LU6.1 session, information can be reliably passed between systems.

LU6.2. The LU6.2 is a newer variation of an LU6. As is true of the LU6.1, the LU6.2 provides peer functions. These functions provide a more flexible foundation for peer communication than is exhibited by LU6.1.

LU6.2 resources form the basis for APPN and HPR; without LU6.2 resources, APPN or HPR could not exist. All of the CP-CP sessions between APPN nodes use LU6.2 sessions.

1.3.5 Sessions

The basic communication paradigm of SNA is the *session*. Almost all communication in SNA is based on a session between two points within the network.

There are four types of sessions:

- Host to host (SSCP-SSCP)
- Host to PU (SSCP-PU)
- Host to LU (SSCP-LU)
- LU to LU (LU-LU)

Each of these has different capabilities and provides for the transportation of different information. Figure 1.4 shows these sessions and how they fit within the network.

1.3.5.1 SSCP-SSCP session. The SSCP-SSCP session is known as a cross-domain connection. This is used predominantly when LUs from

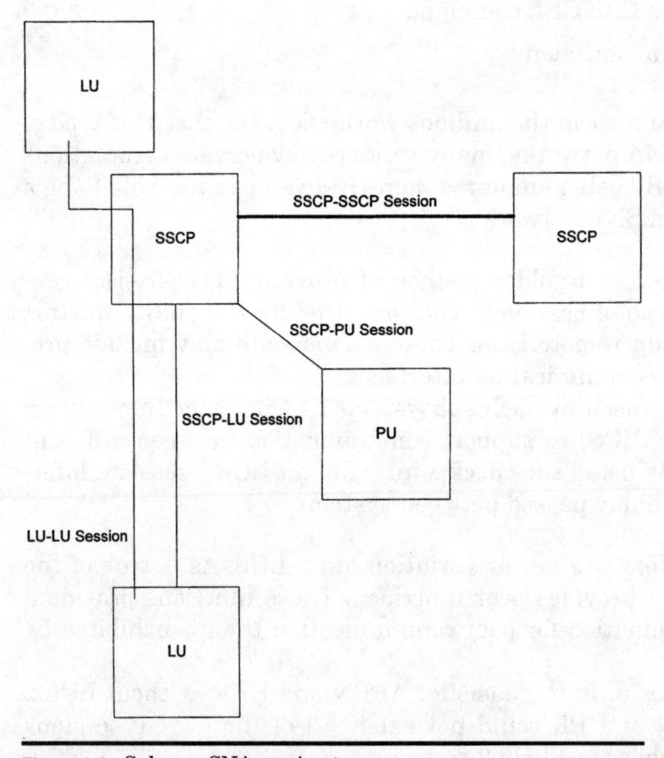

Figure 1.4 Subarea SNA session types.

different domains require communication. The SSCPs communicate, first, to ensure that the destination LU is active and available and, secondly, in order to establish a session.

This type of session is also used for cross-domain network management and for the establishment of the links between those domains.

1.3.5.2 SSCP-PU session. This session is used for the management of a PU and of the LUs associated with the PU. This session is required to be active before the other session types can be activated. The hierarchy of sessions is an important issue throughout SNA networking.

1.3.5.3 SSCP-LU session. The SSCP-LU session establishes connectivity to an LU. This session is required before the LU-LU session can be activated.

Optionally associated with the startup of this session is the passing of information that allows the SSCP to gain dynamic knowledge of the LU and its capabilities.

1.3.5.4 LU-LU session. The LU-LU session is used for the actual application communication. This is the one that is used to communicate with applications such as Time Sharing Option (TSO) and CICS.

1.3.6 Routing

Subarea SNA was designed within the paradigm of the centristic host. All data traffic was between dumb terminals and centralized host processors. Because of this centralized view and the lack of intelligent processing within the network, the route that a session took was based on a determination made by the host. Thus, subarea SNA was based on source routing.

Source routing has advantages. In the network view of all sessions being destined for a host, it seemed to make sense to allow the host to determine the route of a session. It also provided a great deal of simplicity to route determination. If all decisions are made at a single point within the network, the complex routing decisions are not a burden upon the rest of the network.

Source routing also provides for very quick message movement within the network. If a session follows a predetermined path, each node within the network can establish very efficient movement of traffic through intermediary nodes. These nodes create tables that allow traffic to be almost immediately passed through the node. As there is no determination of session path, a "hard-coded" path allows incoming messages to be quickly scanned for the destination address and immediately passed to the correct outbound queue.

By precreating all session routes between nodes, the cost of route determination is almost entirely paid up front. Little processing is incurred to find a route from point A to point B; all of this has already been expended. As all sessions are with the host, it makes sense to locate the determination of a route at this location.

The complexity of intermediary nodes is greatly reduced as routes are preloaded. The intermediary node knows that particular session traffic is always passed through a certain output queue. The tables are already available and the only processing is on session activation to extract the particular route. Once this is done, the intermediary node knows that session traffic with destination Y is always passed to output queue Z. Figure 1.5 shows how to accomplish such a result.

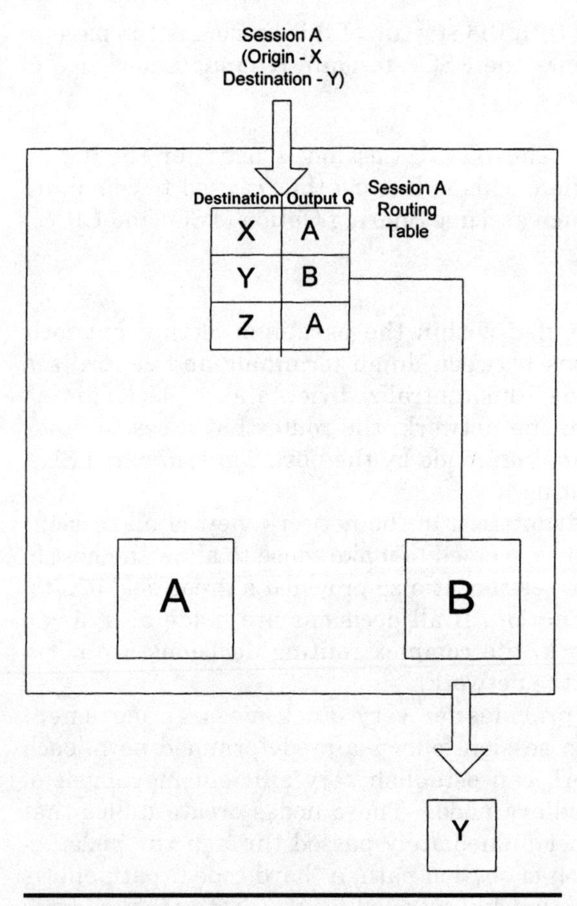

Figure 1.5 Routing within subarea SNA.

1.4 SNA Formats

SNA defines a set of formats that information must adhere to, to enable communication to take place. These formats are further grouped into layers that together enable effective communication.

1.4.1 Components of SNA interfaces

The SNA formats are composed of several components that are reliant upon one another. These pieces were created to enable an orderly expansion of the SNA formats.[10]

The formats can be broken into groups of:

- Transmission header
- Request/response header
- Request unit[11]

Table 1.1 shows the size of each field and their values.

TABLE 1.1 SDLC Frame Format

Field	Size in bytes	Value
Flag	1	X′7E′
Address	1	Poll address
Control	1	Defines purpose of the frame
TH	2, 6, 10, or 26	Provides information to the peer level about the type of information that is contained in the message
RH	3	This format defines the dichotomy of the request/response. This may not be contained in the message.
User data	Variable	The information that needs to be transported
CRC	32 bits (2 bytes)	Cyclic redundancy check
Flag	1	X′7E′

[10]Close observers of SNA have seen expansions of the protocol over the years. Often these expansions are done to enable a facility that was not foreseen.

[11]It is often felt that SDLC (synchronous data link control) is synonymous with SNA, but SNA is actually not reliant on this data link control. This is why I do not discuss the datalink control protocol.

1.4.2 Message format

SNA messages consist of five major components. These are:

1. Link header
2. Transmission header
3. Request/response header
4. Request unit
5. Link trailer

1.4.2.1 Link header. The actual link header is dependent on the data link control (DLC) that is being used. For example, if an SDLC DLC is used, the link header consists of three bytes. For other DLCs, the size and contents of the link header varies.

1.4.2.2 Transmission header. The transmission header (TH) directly follows the link header. For example, when using an SDLC DLC, the TH directly follows the flag (x′7e′), address, and control bytes that define the SDLC link header. Figure 1.6 shows the sequence of headers.

There are two basic types of THs. These are:

- Format identifier 2 (FID2)
- Format identifier 4 (FID4)

The FID2 is 6 bytes long and is used for most end device communication. This TH contains sufficient information to be passed between LUs to create a reliable communication path. Some of the information contained within the TH includes communication parameters and a sequence number for each request so that missing information can be detected.

The FID4 is 26 bytes long and is used for cross-domain communication. Most of the fields in the FID2 are also in the FID4, but often in expanded form. For example, the 32-bit network address is expanded in the FID4 into two 32-bit fields; 32 bits for the subarea number and 32 bits for the resource element.

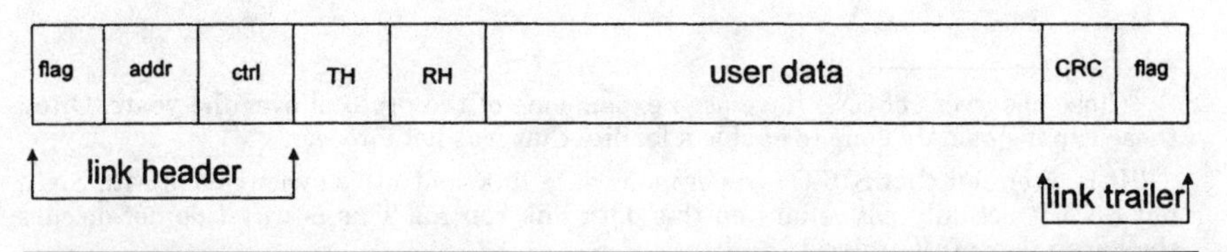

Figure 1.6 Basic SNA data format.

In addition, the FID4 has fields that support a multilink transmission group (MLTG). An MLTG can only be used for cross-domain links. The purpose is to enable support for a logical grouping of physical lines. The link station at each end is responsible for both the queuing of data segments onto the associated links and for the reassembly of these separate segments into a whole message.

1.4.2.3 Request/response header. The request/response header (RH) is a 3-byte field that contains information about whether the frame is a request or response. It also contains information on the current state of the session and change requests to the session status.

1.4.2.4 Request unit. The request unit is the actual data that needs to be transported. This is a variable-length area.

1.4.2.5 Link trailer. Just as is true of the link header, the link trailer is dependent on the actual DLC used on the link.

1.5 Communication Software

When people think of SNA, they have a vision of a large IBM mainframe. IBM, and others, have spent the last 20 years developing software that provides SNA functionality on this type of computer. But three pieces of software, all provided by IBM, symbolize SNA in the central processing location. The software components are:

- VTAM
- NCP
- NetView

Together, these softwares define SNA for most people. Figure 1.7 shows how these softwares fit within an SNA network.

1.5.1 VTAM

VTAM is extremely important in an SNA network. This software, which operates under all IBM mainframe operating systems, provides the SSCP function for an SNA network. Because it represents an SSCP, VTAM is one of the master "cops" of the SNA network.

On a mainframe that uses SNA access methods, VTAM is the interface for all applications. Every request and response passes through the very wide arms of VTAM. As such, VTAM creates the SNA envi-

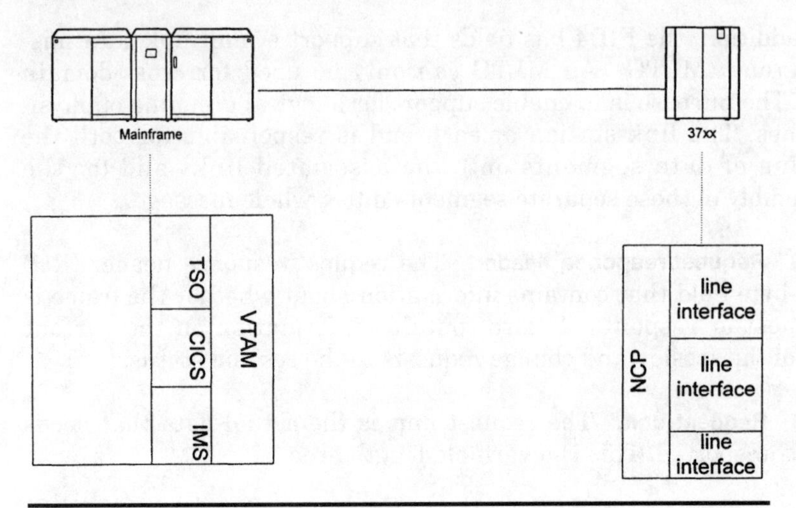

Figure 1.7 Subarea SNA communication.

ronment for the host-based applications, such as CICS and IMS (Information Management System).

As the SSCP, VTAM is also the software that allows LU-LU sessions to be established. All host-based applications interface through VTAM to obtain connectivity to the network of terminals for end-users.

1.5.2 NCP

NCP is the software that operates in the communication processor, such as the 3745. By being outside of the host processor, NCP allows offloading of network functions. These functions include polling of devices, line protocol enforcement, and line error recovery.

The NCP operates as a PU4 within the subarea SNA network. This allows the NCP to also participate in routing decisions and network management. Routing tables are coordinated between VTAM and NCP and are utilized to allow sessions to be established along appropriate paths.

1.5.3 NetView

NetView is the mainframe-based network management system for SNA devices.[12] It uses the communications network management

[12]There are versions of NetView for other environments, but here we will only deal with the host-based product. It is also true that NetView can be made to manage non-SNA resources, but this is also not addressed in this discussion.

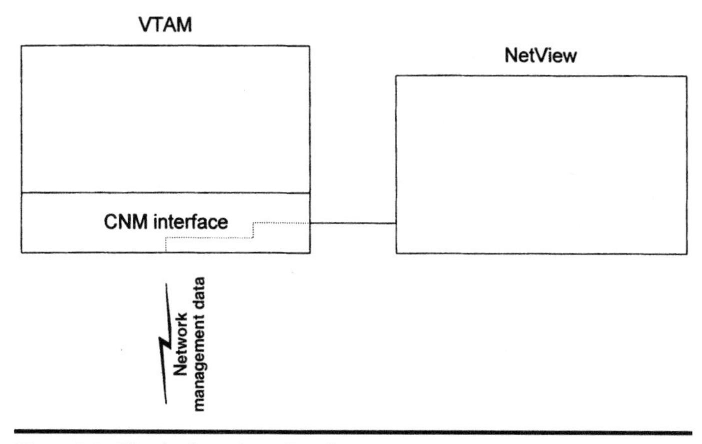

Figure 1.8 Single-domain network.

(CNM) interface of VTAM to obtain access into SNA management. Figure 1.8 shows the relationship between VTAM and NetView. NetView uses the CNM interface of VTAM to obtain unsolicited network management data.

NetView is able to obtain both solicited and unsolicited network management data. It is also tied into the operating system to allow operating system commands to be issued.

By providing the breadth of access from a single point, NetView can provide a robust management point for SNA networks.

1.6 SNA Networks

SNA networks range from the simple to the complex. Though I will not go into excruciating detail, it is important to get an appreciation of the types of SNA networks. For this purpose, we will look at two levels of complexity:

- Single domain
- Multiple domain

1.6.1 Single domain

The single-domain SNA network contains only one SSCP. All of the resources within the network are owned by the one SSCP and thus belong within its domain.

Figure 1.9 shows a single-domain network. Though this network is single domain, it consists of several communication controllers and

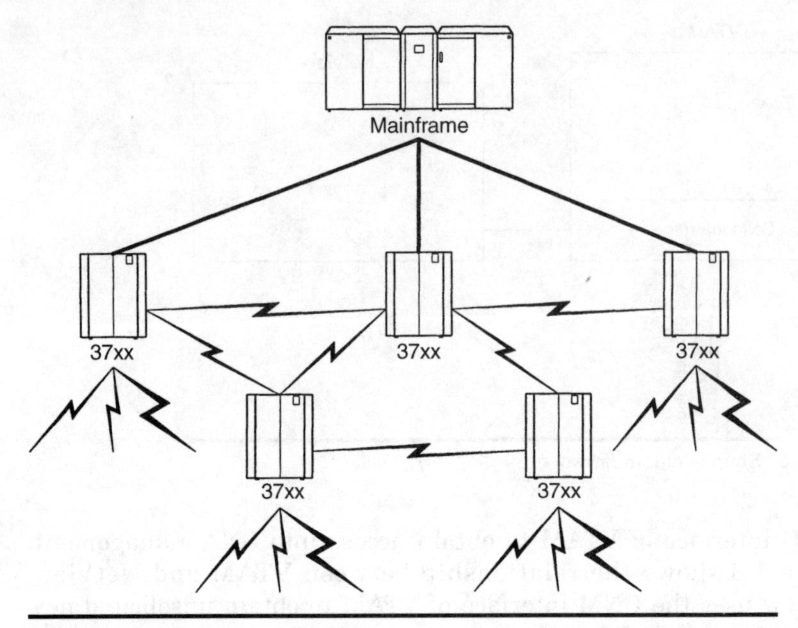

Figure 1.9 Single-domain network.

many devices. This is not a simple configuration, even though it is a single-domain network.

In the case of this example network, all resources in the network[13] are owned by the single VTAM. The number of total resources could be very large and management of such a network can be a challenge.

1.6.2 Multiple domain

A multiple-domain SNA network consists of more than one SSCP. The addition of more SSCPs provides many benefits, such as load distribution and host backup and recovery, but it can also greatly complicate the execution of a network design, can make network management more difficult, and may require a much more complex routing scheme.

Figure 1.10 shows a multiple-domain network. The communication controllers are owned by both hosts, as are the lines that are attached to the commonly owned front ends.

[13]The resources include communication controllers, lines, PUs, and LUs.

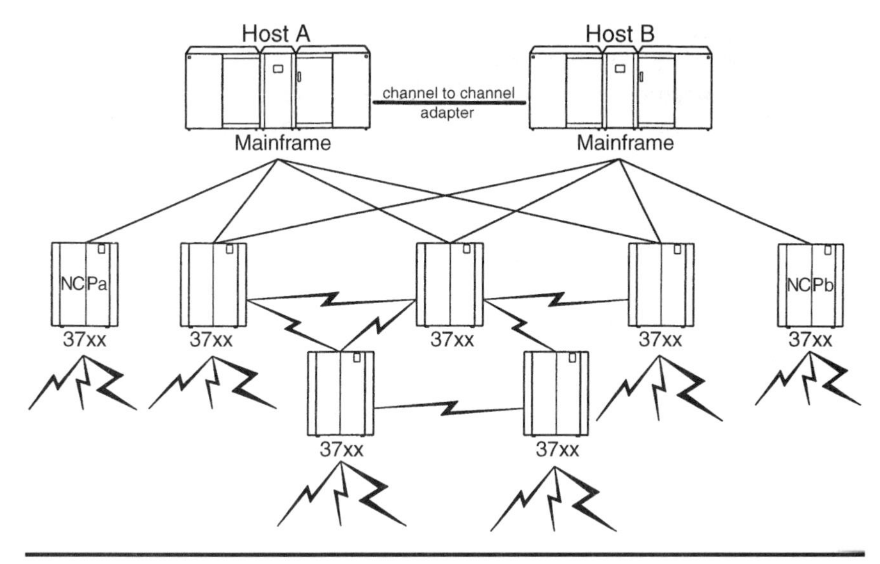

Figure 1.10 Complex multidomain network.

Note that there are also two communication controllers that are not owned by both hosts (NCPa and NCPb). These front ends are owned by a single host, thus there is no capability of these to communicate with the other host without first going through the owned host. Thus, NPCa cannot communicate with a resource on Host B without going through Host A. If Host A is not available, no host communication is possible, even if Host B is available.

The ownership and backup configurations can be much more complex. It is possible, in this shared environment, for certain resources to be owned by one host, even though they may appear to be shared. For example, though it may not make sense, one of the common NCPs may be owned by only one of the hosts.

1.7 Summary

We have seen the background of subarea SNA. The reason that IBM designed SNA was both for the customer and to serve self-interest. Customer networks were starting to get quite large and complex and IBM saw that the desire for dramatically increasing the size of those networks was also increasing.

At the same time, IBM was wasting a great deal of time and money in duplicated efforts to develop networking software. As the communication interface resided in each subsystem, the amount of duplication was extensive. In addition, as the code had to be developed for each product, the accumulated cost to IBM was becoming sizeable.

In an effort to reduce this cost and ensure a consistent interface, along with the requirements from their customers, IBM designed and developed SNA.

The architecture designated types of interfaces and how they interoperate. The creation of a consistent, architected interface allowed development of communication components independently while increasing the probability that these components would work when combined.

Chapter

2

Advanced Peer-to-Peer Networking (APPN)

As subarea SNA networks grew in size and companies grew more dependent on network availability, subarea SNA networks started to show some deficiencies. These included the ability to reduce the support requirements. The capabilities of subarea networks to provide a smooth path to larger networks became difficult. Support for the network grew exponentially as the number of nodes increased and resources become more mobile.

APPN is an architecture for the enhanced use of dynamic topology and directory services. Unlike subarea SNA, APPN allows for dynamic address resolution, route determination, and LU registration. The static routes and LU definitions of subarea SNA are replaced by the dynamic topology and route determination of APPN.

2.1 History

APPN was first introduced to the public in 1985. The architecture evolved from several "problems" of subarea SNA. These include:

- Ease of use
- Dynamic route determination
- Dynamic topology

Subarea SNA used static routing and definition tables. These tables can be updated dynamically through some cumbersome methods, but even these changes are only temporary. In most cases, these dynamic definition changes are not kept across system startups. This resulted in cumbersome disaster recovery plans.

In addition, local area networks (LANs) and dynamic resource allocation in the host expanded in availability. These capabilities resulted in resources that were not static in location. Subarea SNA was not able to support these configurations efficiently.

Many customers of subarea SNA had reached their limits in attempting to provide a fail-safe environment for connecting their sites. Customers had created and used many difficult solutions that worked, however these were pieced together in such complex ways that it was a constant challenge to keep the parts operational. The continual struggle to support dynamic address resolution in an architecture that did not natively support them lead to many precarious solutions.

Requester/server solutions also started expanding past the test phase. These application pairs brought to light several strong requirements. These included:

- The ability to reach a random destination without the need to predefine the path between endpoints
- The ability to move where an application was operating and the network being able to dynamically find the new location

These requirements lead to the need to extend subarea SNA. The question became one of whether the expansion would be evolutionary or revolutionary—would the new design build directly on subarea SNA or would it start upon a new base? This was not a simple answer since IBM had invested millions of dollars in the subarea SNA base. In addition, IBM customers had invested billions of dollars in this solution.

A decision was made to start with a new base, but to use the knowledge gained through subarea SNA to create an architecture that would provide solutions for today, but also provide a basis for the networks of tomorrow.

The requirements were the following:

- *Distributed directory searches.* The objective was to use an architected method of obtaining directory information. The requirement was to reduce the amount of information that must be predefined and coordinated between multiple locations.

- *Topology and route selection.* Rather than use the static routing tables that exist within subarea SNA, it would be better to let the network make these decisions. The network is capable of these decisions because of the high processing power possessed by the nodes within the network.

- *Adaptive pacing.* Subarea SNA used a static pacing value. Though this was successful in reducing the probability of network congestion, it was not able to dynamically adjust to network conditions.
- *Transmission priority and class of service support similar to subarea SNA.* Subarea SNA was successful in creating a method to create a prioritization scheme that was more robust and more fully implemented than any network architectures. APPN must meet these same objectives without imposing new requirements on the customer.

2.2 APPN Architecture

APPN is a structured communications architecture. The components of an APPN node are described and the interface between components are defined in order to provide a method of creating an APPN node. Because it was built upon a new base, it was isolated from the architectural problems of subarea SNA.

2.2.1 APPN node types

APPN architecture has defined three types of nodes. Each of these types provide a different level of support for the APPN architecture and was defined to allow a migration from subarea SNA.

The three nodes types are:

- Low-entry networking (LEN)
- End node (EN)
- Network node (NN)

2.2.1.1 Low-entry networking (LEN). A LEN node is a migration path from the subarea SNA network architecture. This node type implements a PU 2.1 appearance to the host to allow dynamic LU allocation, but does not support the directory services of APPN. As a result, the LUs on a LEN node must be predefined to the host.

This node allows APPN nodes to be connected to a subarea host, while not requiring the host to implement any APPN logic. This migratory path allows the subarea SNA host to connect to APPN nodes as LEN. Although definitions are required for this connection, the LEN node may actually be a gateway into the APPN network. This scenario is shown in Fig. 2.1.

The AS/400 operates as a gateway from the AS/400 APPN network into the subarea SNA network downstream from the 37x5. This is the same migration path that IBM created in VTAM V3R3.

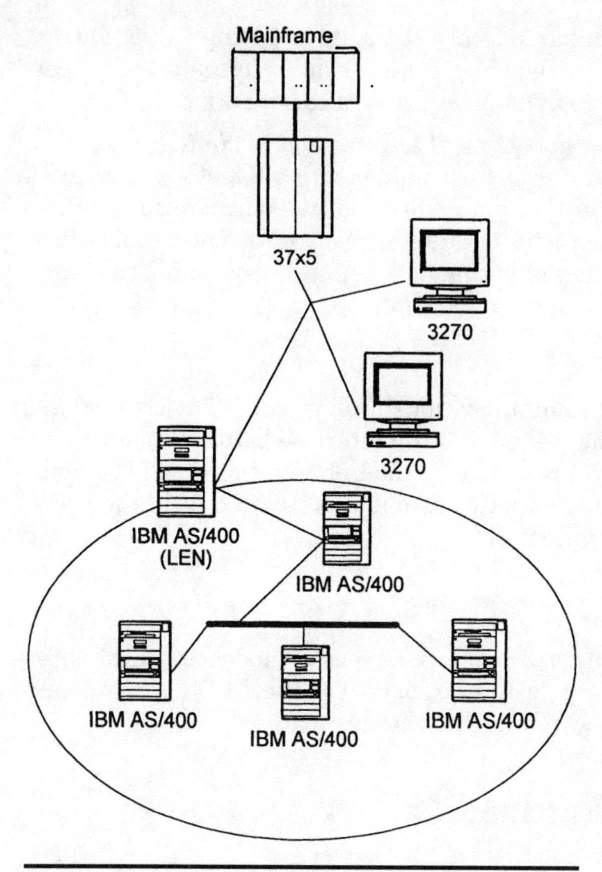

Figure 2.1 APPN network of AS/400 with LEN gateway.

When going through a LEN gateway, the subarea host must have definitions for every resource that needs to be reached in the APPN network. Each of these definitions points to the LEN gateway as the path to the resource. Once a request is passed to the gateway node, this node uses APPN logic to continue the route to the actual destination.

Figures 2.2 and 2.3 illustrate a LEN gateway configuration. The first shows the physical connectivity and the location of LUs within the network. This is not a complex network, but a common configuration of APPN nodes. Figure 2.3 shows the logical view of the network, from the point of view of the host. This figure shows that the details of the configuration "disappear" in this logical view. This is a result of ability of the host to depict the physical connectivity through the APPN network. The gateway hides the actual network by contracting

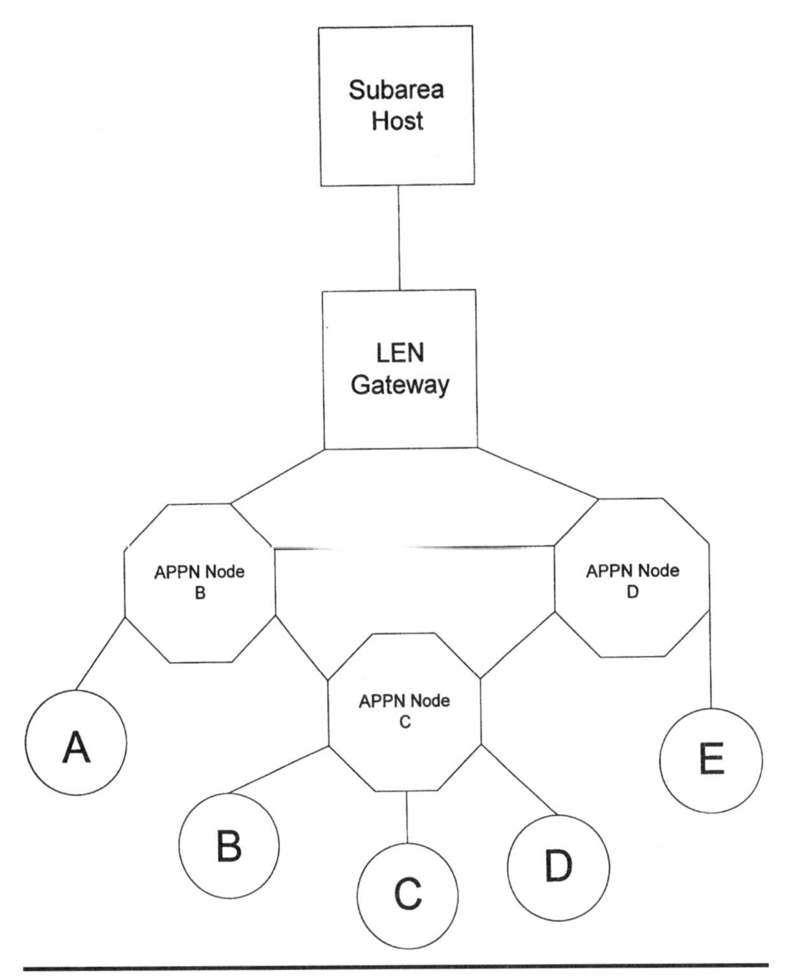

Figure 2.2 LEN gateway configuration—physical.

the APPN network into the view of being directly connected to this node, instead of the actual connectivity.

Network management in this configuration can be difficult, because the details of the configuration can be hidden. If the network control technician is provided the view in Fig. 2.3, it can be difficult to find a failure between the gateway and the actual session LU. Without this detail, it is more difficult to pinpoint a failure.

2.2.1.2 End node (EN). The EN is a true APPN node because it supports many of the APPN features and fully participates in APPN networks.

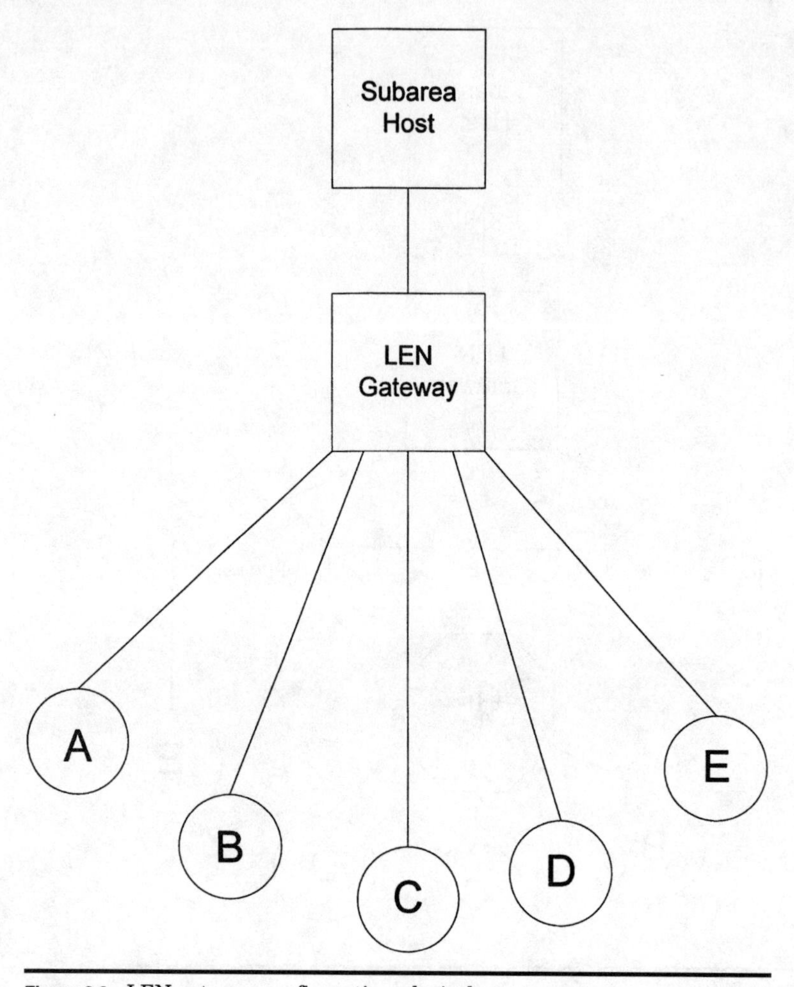

Figure 2.3 LEN gateway configuration—logical.

The EN, as the name implies, is at the outer perimeter of an APPN network. Figure 2.4 illustrates how an EN resides in an APPN network.[1] The application LUs are normally located on the ENs. For this reason, VTAM mainframes often migrate into ENs when migrating from subarea to APPN.[2]

---

[1]Note that there are normally many more ENs than NNs.

[2]A discussion of the migration path from subarea SNA to APPN appears in Chap. 7.

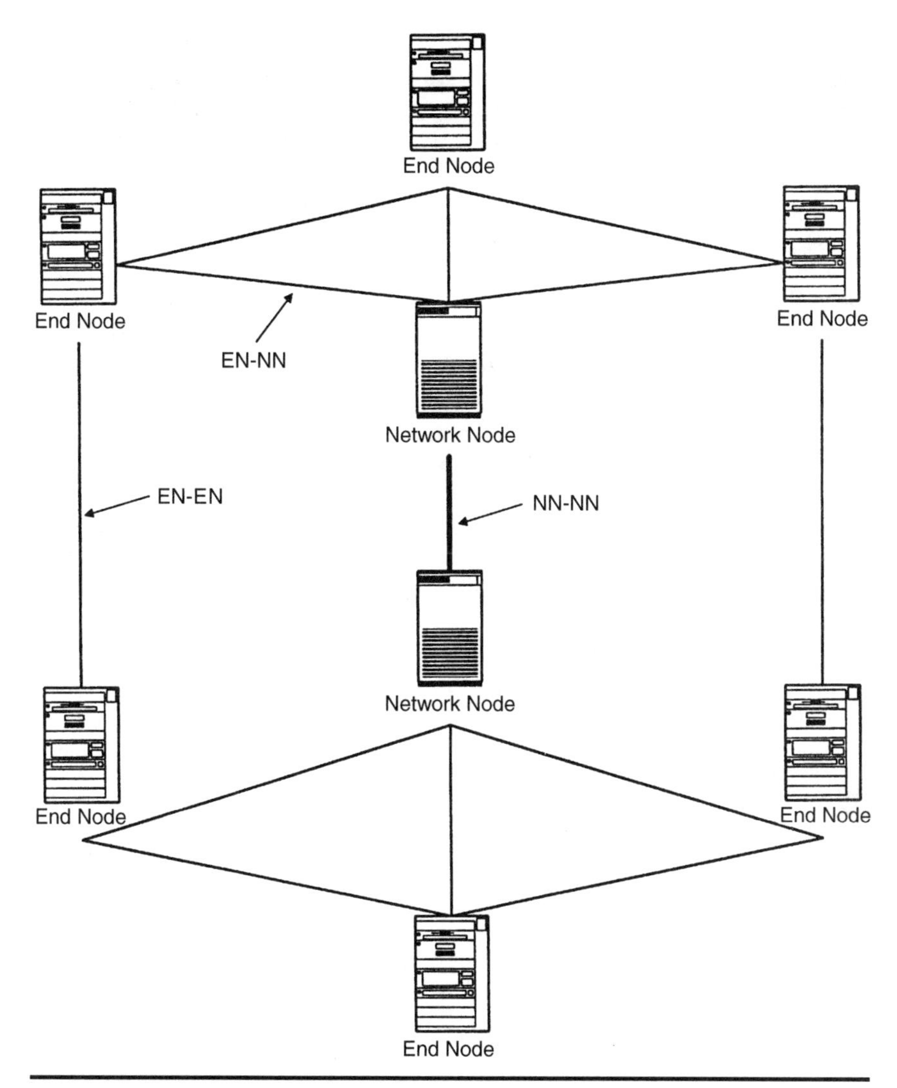

Figure 2.4 End node in APPN network.

The EN provides all APPN services except for routing and topology; these functions are those of a NN. The EN does provide such functions as:

- Control point–control point (CP–CP) session
- Participant in directory searches

By not providing full directory services and any routing and route calculation capabilities, the overhead of these tasks, as implemented on a NN, is reduced. This extra capacity is thus available to be utilized by the processing requirements of the associated LUs and their related transaction programs.

2.2.1.3 Network node (NN). The NN is the heart of an APPN network. The enhanced services of an APPN network are realized through the NN.

The NN provides all of the services of an EN but includes routing, topology, and full directory services. It is through these services that the dynamic benefits of APPN are gained. The NN provides support for affiliated ENs.

Network nodes may optionally support LUs. These LUs gain no special services by being attached to an NN, except for a potential performance gain in obtaining directory services locally. This gain must be countered by the overhead associated with servicing the LU.

2.2.2 APPN communications interface

APPN nodes use LU 6.2 sessions between themselves. These sessions are not special unto themselves; they use the normal LU 6.2 session interface to communicate. Thus, the session layer of APPN does not require any new development.

The new factor is the communication between the LUs in each node. These LUs use extensions of the normal data flows of such LU 6.2-capable systems such as CICS and OS/2 Communication Manager (CM). The external differences of an APPN node exist at the application layer.

APPN nodes support the architected transaction programs for APPN CP–CP sessions. These transaction programs provide the additional logic necessary for this protocol.

The CP–CP session is the foundation of APPN. Through these sessions,[3] nodes gain awareness of their neighbors and, from them, the rest of the APPN network. Indicators passed during the session start-up determine the LU capabilities.[4]

---

[3]There are always two CP-CP sessions between a pair of APPN nodes. Each of these session is unidirectional and is persistent.

[4]These indicators are coincidentally called an LU Capabilities vector.

2.2.3 Node structure

APPN nodes use a strict node structure to provide services to other LUs. Figure 2.5 illustrates the base node structure. This figure delineates the major components of an APPN node. These are:

- Node operator facility (NOF)
- Control point (CP)
- Intermediate session routing (ISR)
- Logical unit (LU)
- Path control (PC)
- Data link control (DLC)

Although some of these components are similar to those within subarea SNA, these functions have been further architected and refined.

2.2.3.1 Node operator facility (NOF). The NOF component of an APPN node provides operational control of the node. This interface can be used either by a person or a program.

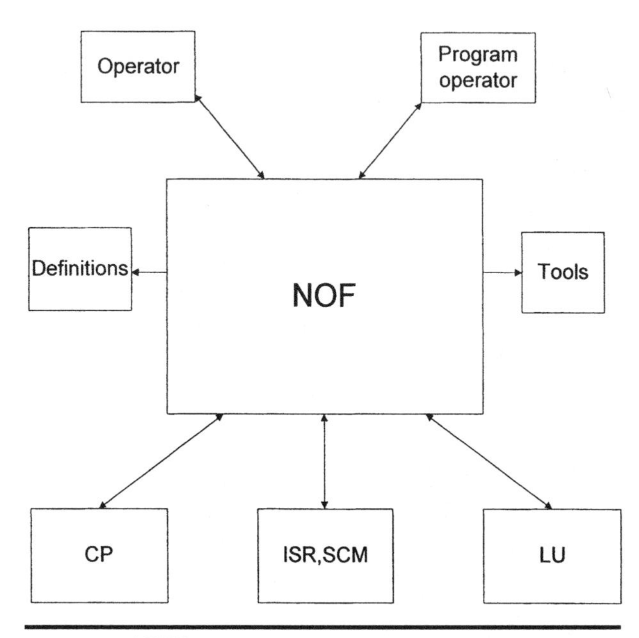

Figure 2.5 APPN components.

The NOF component provides initialization services for the node. This allows for the activation of the other node components. NOF is also the location for all definitions. The definitions include:

- Node characteristics
- Local link stations
- Local, and possibly remote, LUs
- Adjacent nodes
- Transaction programs
- Diagnostic tools

2.2.3.2 Control point (CP). The control point (CP) is the heart of the APPN node and provides the actual management and interface to other APPN nodes. Although the LU (see Section 2.2.3.4) provides the underlying path, the CP is where the APPN protocol is actually implemented.

The CP is composed of subcomponents that are discussed later in this chapter. These subcomponents are where the APPN architecture is implemented. The layered method of software implementation and development is used within the node by architecting the control point as a set of subcomponents.

2.2.3.3 Intermediate session routing (ISR). The ISR component of the APPN node provides the ability to route traffic from one link to another. This component is limited only to NNs; the EN does not provide any routing of messages.

ISR is a key component for users, as most users want to be able to route messages through their networks. The dynamic routing that APPN is capable of is already used by many customers to ease the support requirements of their networks. By allowing the network to locate resources dynamically, definitions are not only reduced, thus saving critical human resources, but also resources can be moved on a real-time basis to adjust to changes in resource demands.

2.2.3.4 Logical unit (LU). The LU facilitates the actual session between APPN nodes. Session initiation and maintenance are supported by this component of the APPN node. This portion of the APPN node is a slight extension of the LU of the subarea SNA node by possessing additional transaction programs. The subcomponents include:

- Service transaction programs
- Session manager

- Resource manager
- Presentation services
- Half-session manager

Together, the LU provides the underlying foundation for the session between APPN nodes.

2.2.3.5 Path control (PC). Path control consists of manager and element components. The manager subcomponent of the APPN node establishes and detaches connections to other APPN nodes. The PC passes knowledge of other nodes to several components so they have current knowledge of the APPN network structure.

The PC element routes messages to other components of the node and provides support for segment generation and some message error checking. This is also where prioritization of message traffic is actually implemented.

2.2.3.6 Data link control (DLC). This component is synonymous with the DLC layer of the ISO model. Message transmission and reception is done in this component. Whatever headers are required to allow different DLCs to communicate is implemented in this component. Thus, the higher components are isolated from the intricacies of DLC implementation and message modification to fulfill the requirements of a specific DLC.[5]

2.2.4 APPN method of support

The APPN node consists of several components. As discussed in Sec. 2.2.3, the structure of the node provides architected interfaces between components. This facilitates more efficient node development, because of the lack of a single point of reference. The node is provided with a more complete description of the node and the individual logical components.

It must be stressed that Sec. 2.2.3 illustrates a logical structure that does not have to be followed; this is a logical abstraction of the node that is not apparent outside of the node itself, thus, there is actually no requirement to follow this node structure.

The present section describes the actual interaction between APPN nodes and the services that are provided to components logically above the APPN components. This includes:

---

[5]This is a good example of the advantages of a layered architecture. The higher levels are unaware of the specifics of a DLC. Thus, the DLC can be changed without any impact to the higher levels.

- Dynamic directory services
- Topology and route selection
- Adaptive pacing
- Transmission priority/class of service

2.2.4.1 Dynamic directory services. Full APPN nodes provide a dynamic directory service to supported LUs. This directory is fully implemented within the NN and partially implemented by an EN. In either case, dynamic directory services within the APPN node dynamically finds a path to the specified destination.

When an LU requests connection to an LU, it sends an architected request that results in a set of events to locate the specified destination. Such a locate is done whenever a request for session services is received, such as when an LU requires a session with a destination.

By allowing for a dynamic directory, the location of a destination does not have to be preconfigured. Instead, the network locates session destinations dynamically.

Both the NNs and ENs have some form of directory services. The EN provides this as a local directory. This directory contains the local LUs and the lines (transmission groups [TGs]) that are available for use by sessions.

Figure 2.6 depicts a typical APPN network. EN1 has a network node server (NNS) of NNa. If a session setup request for LUb arrives at ENa and ENa does not already know where LUb is located, ENa sends a LOCATE request to its NNS, NNa. NNa broadcasts the LOCATE throughout the network. When a response is returned, NNa gains knowledge of where LUb is located. NNa updates its directory cache and ENa updates its tables. In this way, the search and route calculation, which is costly in time and processor cycles, is eliminated for the next session request to the same destination. Instead, a directed search is sent to ensure that LUb is still at the same location. Assuming that this is true, the cached route is utilized.

2.2.4.2 Topology and route selection. Rather than use the static routing tables that exist within subarea SNA, it is better to use a dynamic method. With the level of processing power available within the network and the high cost of support staff,[6] by having the components

---

[6]Because the human resources costs required to create the routes that exist in subarea SNA are high, such costs can be reduced by almost eliminating route configuration.

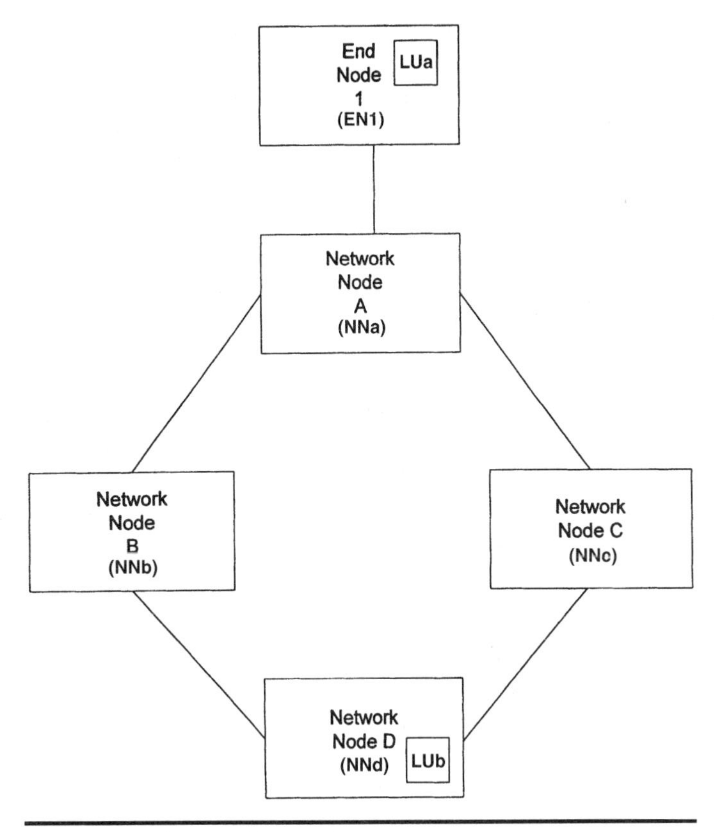

Figure 2.6 Typical APPN network.

within the network dynamically find resources, companies cannot only provide better service, but can also save money.

An APPN node requires a limited predefinition. The minimum requirements necessary are to define the name of the node, the name of the CP, and the physical configuration of the output ports.

The name of the node and the CP cannot be dynamically determined by anything within the node itself; it must be defined. Likewise, the physical ports must be defined to the node. This definition includes the address of the port, the DLC that will be used, and other physical configurations. It is not necessary to configure the physical components that are downline from the port. This information is learned dynamically upon contact with the remote link station.

Route calculation is also done dynamically; or at least, it is not predefined. Instead, routes are calculated by using either an existing directory cached entry or by calculating the route using the current knowledge of the network and its connections.

Assuming that a route has not already been calculated by an NN, an NN will use the current network topology and directory information and calculate a route between point A and point B. Once calculated, the information is cached for later use.

Looking at Fig. 2.6, ENa first searches its cached information to find a route to LUb. Assuming that this information does not already exist, NNa (ENa's network node server) goes through a similar determination. If it does not know where LUb is located, it sends a request to all of the NNs that it has a session. Eventually, the location is returned from NNd.

NNa calculates a route to NNd from ENa. This route is independent of how the request arrives, either as NNd or as the sessions that NNa and NNd have. Instead, all paths that ENa and NNd have is used by NNa to calculate this route.

Again, assuming that NNa has calculated a route from ENa to NNd for the network shown in Fig. 2.6, a route may be calculated as follows:

$$\text{ENa} \rightarrow \text{NNa} \rightarrow \text{NNb} \rightarrow \text{NNd}$$

If a link became active from ENa to NNd (see Fig. 2.7), this information is not reflected in the cached information. When ENa looks in its cache, the above route is located, instead of the more potentially more optimal route of ENa→NNd. Since LUb is still at NNd, the cached information will be accurate, but not the best choice.

It can thus be understood that the path located is not necessarily the currently *best* path. Network nodes determine the best path to a destination. This path is cached so that later path requests to the same destination are immediately available. It is very possible that a new and better path has become available, but the APPN node will not locate this new path, but will instead return the cached path that was previously calculated.

This cached entry will eventually age out, but this can take as long as days, which is not an optimal duration. A path also becomes aged upon the failure of a component along the route. Thus, it is possible to obtain a new path by forcing a link to become inactive. When the NN gains this awareness and a new route request arrives for the destination using that path, a new route will be calculated.

Although this scenario is not normal, it should be understood, so we can appreciate how APPN operates and to see its shortcomings. These sessions are established, in the previous case, but will not be "optimal" paths.

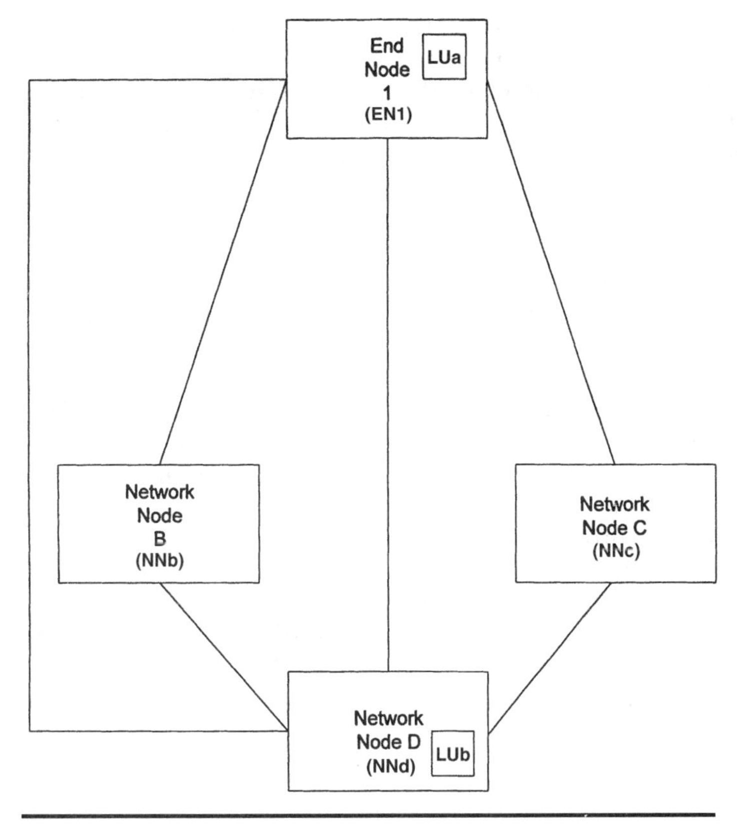

Figure 2.7

2.2.4.3 Adaptive pacing. Subarea SNA used static pacing values to reduce the probability of network congestion. However, it was not able to dynamically adjust to network conditions. For example, if a session does not have any competition for the link, throughput may be very high. But if someone in the network decides to transfer a large file along your link, competition for the link is dramatically increased. As a result, although your static pacing value may have worked under low volume conditions, under stress conditions, it may not fulfill its objective.

APPN has a patented adaptive pacing algorithm. This algorithm allows for the pacing window to dynamically adjust to the current network utilization and to the capacity within the network. As utilization increases without a corresponding increase in contention for the network components, the pacing window is increased. Likewise, if too much queuing is exhibited within the network, pacing windows are decreased until the queues diminish.

This adaptive pacing allows APPN networks to exhibit enhanced throughput characteristics under variable network conditions. It also provides better network utilization by allowing components within the network to vary their utilization.

2.2.4.4 Transmission priority/class of service. One of the strong points of subarea SNA is its transmission prioritization scheme. This provides for several levels of prioritization that can result in satisfactory throughput for different types of data. Thus, for example, interactive traffic can be given priority over file transfer traffic.

Class of service fulfills the different service requirements of different types of traffic. One session may require a high priority, but you do not want this session unless it can be encrypted. Other sessions may require a path that can provide a great deal of bandwidth. Each of these session requirements can be specified through the use of the class of service (COS) table.

Transmission priority allows for the differential queuing of data to physical connections. By allowing for different queues, data can be serviced for transmission in an order other than first-in, first-out.

Subarea SNA and APPN use four transmission priorities. These are:

- Low
- Medium
- High
- Network

The first three can be specified within the COS table. The last, network priority, is specified only within reserved COS table entries.

Unfortunately, the COS is resolved during session setup by the NNS of the requesting LU. Once a session is established, its session traffic cannot traverse a different path dynamically.[7] The session flow can only be rerouted by terminating the existing session and requesting a new one. At that point, the NNS will recalculate the session path utilizing the current topology database.

---

[7]High-performance routing (HPR), an extension of APPN, does provide for this dynamic session routing. This rerouting is normally used when a session segment becomes unavailable. In this case, the session traffic can be dynamically rerouted to a new path without the knowledge of the session endpoints. See chapter 3 for more information.

2.3 New SNA Formats

In order to provide additional services, APPN nodes use some new SNA commands and formats. We will review some of these commands and formats and how they assist the APPN node in providing the extended services.

2.3.1 XID3

During the initial contact between APPN nodes, they exchange initial information. This information is transmitted through the use of an extension to the XID command. Though an XID is not a new command, the APPN node uses an extended form known as XID format 3, or XID3. This XID provides a great deal of information about the node's capabilities. If these sessions are desired, the node name and type, and many other pieces of information that allow for the smooth transport of data.

Table 2.1 shows the format of the XID3. It can be seen by reviewing the format that the XID3 provides a great deal of dynamic information on the remote node and the capabilities that the node will exhibit across the link.

One of the reasons for the differences in node capability for different links is to reduce the topology updates that are required throughout the APPN network. The more nodes advertising that they have network node capability, the more overhead exists for those topology updates throughout the APPN network. Limiting the links that exhibit network node capability also can dramatically reduce the overhead associated with route generation. This is because the algorithm grows exponentially as the number of NNs and links increases. The "extra" links probably add nothing to the network availability. As such, it is advantageous to eliminate them from the APPN topology. Since it is possible for a node to dynamically modify its capabilities, if it becomes advantageous to become a network node, it is possible to provide this service.

2.3.2 PU 2.1

All SNA communication is based on the underlying foundation of the PU. In subarea SNA, the main PU type, at the periphery, is PU 2.0, or just PU2. This PU provides the base management services required of its LUs. The base functions provided for the existence of supported LUs and the management of the base upon which these LUs resided.

TABLE 2.1 XID Format

Byte	Bit	Description
0	0–3	Format of XID X´0´ fixed format X´3´ variable format for PU2.1-PU2.1
	4–7	Type of node sending XID X´1´ T1 X´2´ T2 X´4´ subarea node (T4 or T5)
1		Length of variable fields This is reserved for XID0
2–5 |7		Node identification
8–9	0	INIT-SELF support 0—INIT-SELF may be sent 1—INIT-SELF may not be sent
	1	Standalone BIND support 0—BIND may be sent to XID sender 1—BIND may not be sent to XID sender
	2	Whole BIND generated indicator 0—node generates segmented BINDs 1—node does not generate segmented BINDs
	3	Whole BIND receive indicator 0—node can receive segmented BIND 1—node cannot receive segmented BIND
	4–7	reserved
	8	ACTPU suppression indicator 0—ACTPU for SSCP-PU session requested 1—ACTPU for SSCP-PU session not requested
	9	Networking capabilities indicator 0—sender is not a network node 1—sender is a network node
	10	CP services 0—CP services not requested or provided 1—CP services requested or provided When bit 9 = 0, CP services requested. When bit 9 = 1, CP services provided
	11	CP-CP session support 0—CP-CP session not supported on this link 1—CP-CP session supported on this link

TABLE 2.1 (*Continued*)

Byte	Bit	Description
	12–13	XID state indicators 01 negotiation-proceeding exchange 10 prenegotiation exchange 11 nonactivation exchange
	14	Nonactivation exchange secondary-initiated 0—secondary initiated not supported 1—secondary initiated supported
	15	CP name change support 0—sender will fulfill XID exchange, but name cannot change 1—supports name change during nonactivation exchange
10	0	Adaptive BIND pacing support as BIND sender 0—adaptive pacing of BIND not supported 1—adaptive pacing of BIND supported
	1	Adaptive BIND pacing support as BIND receiver 0—not supported 1—supported
	2	Quiesce TG request indicator 0—sender requests that the receiver generate topology update that the TG is operative 1—sender requests that the receiver generate topology update that the TG is quiesced
	3	PU capabilities support 0—PU capabilities vector not supported on ACTPU 1—PU capabilities vector supported on ACTPU
	4	APPN border node support 0—not supported 1—supported
	5	Reserved
	6–7	Qualifier for adaptive BIND pacing 00 adaptive BIND pacing for both dependent and independent LUs and is nonnegotiable 01 adaptive BIND pacing for both dependent and independent LUs, unless overridden by partner 10 reserved 11 adaptive pacing for independent LUs only (retired)
11–14		Reserved

TABLE 2.1 (*Continued*)

Byte	Bit	Description
15	0	Parallel TG support indicator 0—not supported 1—supported
16		Transmission group number
17		DLC type X´01´ SDLC X´02´ System/370 channel
18–*n*		DLC dependent section

Abbreviations: BIND; XID.

The network and transport layers of this foundation were fairly simple. The PU2 was based upon the idea that it managed a single link to all facilities. As such, there was no routing logic within this PU. The DLC had a single path to all destinations; routing was not a consideration.

The idea of a dynamic network topology required the existence of a routing agent within the node. If dynamic topology was to be of any use, it was imperative that the node determine if the multiple paths exist and if the node itself possess more than one physical path. The PU2.1 node was created to fulfill these requirements.

The PU2.1 node (the previous PU2 node became known as PU2.0) has the logic to provide this routing logic. It provides for multiple physical connections to the network. It also supports parallel connections between nodes. This is similar to the MLTG that is available for subarea SNA nodes. However, some restrictions do exist and will be discussed later.

Figure 2.8 portrays a typical network. In this simple network, Node A and Node D have two paths between them. The PU2.1 logic in each must be able to ascertain that these multiple paths exist and allow a routing decision to be made as to which route to use for each session. These decisions are made by taking into consideration the characteristics of each path.

For example, if Path 1 is a T1 link while Path 2 is a 1200-bps dial circuit, it may be advantageous to use Path 1 for large file transfers. At the same time, if Path 2 has an encryption device on it, it would be used for sessions that require this security characteristic, even though the width of the path is less than optimal.

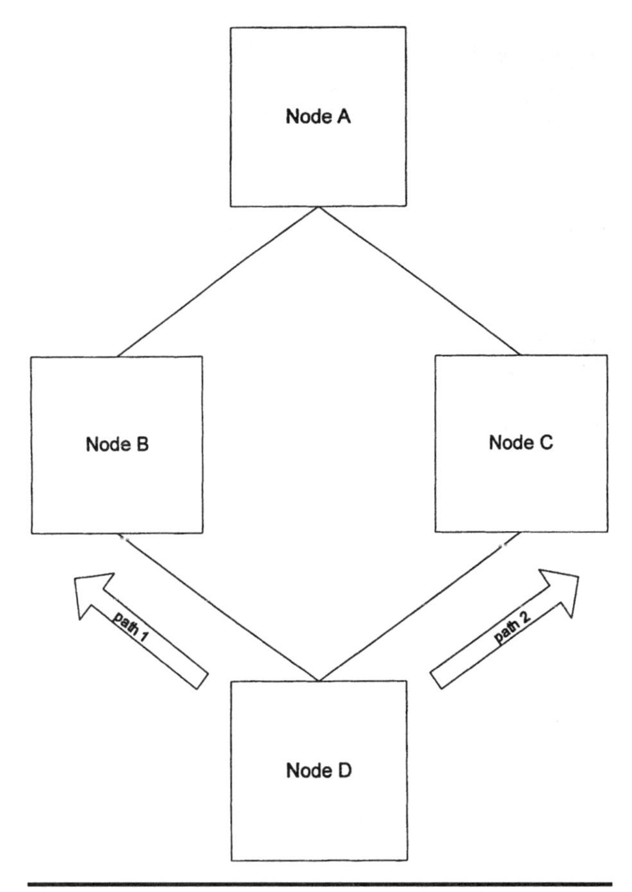

Figure 2.8 Network with multiple physical links.

2.3.3 LOCATE GDS

The LOCATE GDS variable is one of the keys of the APPN network directory services. This GDS variable is the wrapper for the directory searches and responses.

The locate is sent after an LU requests session services to establish a session with a destination. When the node gets the request for services, it scans its cached directory. If the destination is located, a directed locate is sent out to determine if the resource is still in the same location, as shown in Fig. 2.9. The source of the request sends a directed search to the destination. Assuming that the resource is still located there, a positive response is returned. At that point, the node has the knowledge necessary to reach the destination.

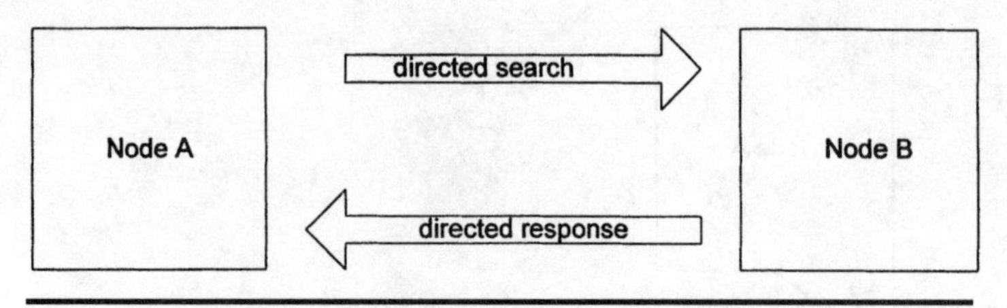

Figure 2.9 Directed locate request/response.

If the resource has moved or if the originating node does not know where the resource is located, a *broadcast locate* is sent. This request is sent throughout the APPN network to learn the location of the resource. There may be several responses to this request; it is the responsibility of the serving network node to cull through the multiple responses and determine the best route to take.[8]

2.4 Communication Software

Several products provide APPN support. These range from IBM communication products to vendor products that can coexist within an APPN network. Because the architecture for APPN has been opened up to vendors outside of IBM, the ability of others to support the architecture is facilitated.

The type of products that support connectivity into and through APPN networks extend from those that are able to coexist to those that fully implement and participate on the APPN network.

2.4.1 AS/400

The AS/400 represents one of the major implementors of APPN. Networks of AS/400s epitomize APPN networks. It is within this arena that APPN has grown and flourished.

The AS/400 has been the real-world laboratory of APPN technology. Users of this product have expanded the horizons of APPN and enabled the architecture to expand past research.

Most AS/400 networks are small when compared to VTAM networks, but they provide many good examples of how APPN can enable reduced configuration and ease of use. These terms are the watch-

[8]It is also possible to extend the search into the subarea portion of the mixed-network topology network.

words for the AS/400; they are also embodied in the network side of this processor.

Though AS/400 networks support APPN, they often need to interface with subarea SNA VTAM networks. Until VTAM provided APPN support in Version 4, the method for interconnection was through low entry networking (LEN) connectivity.

Interconnection was effected by configuring a link for the AS/400 so that it gained a bridge to VTAM. The link was normally through an intermediary NCP. This connection allowed session traffic to flow. Such a link did not provide dynamic definition[9] and parallel links.

2.4.2 VTAM

VTAM first supported APPN with VTAM Version 4. This, along with NCP Version 7, allowed VTAM and NCP to share responsibility to provide an APPN appearance to other nodes. The end result is known as a *composite node*. Figure 2.10 shows an example of this topology. The VTAM and NCP within this figure provide the appearance of a single node to the rest of the APPN network. The distinction between VTAM and NCP disappear within this network.

It is possible for this composite node to be a gateway into the subarea SNA portion of the network. Thus, resources can be migrated into the APPN network, while allowing connectivity to both the APPN and subarea SNA networks. The composite node gateway can support the registration of LUs to the APPN network, while also allowing subarea SNA resources to reach resources on the APPN side. This dual image, as seen from VTAM, is one of the strong migration methods that exist in moving from subarea SNA to APPN.

The APPN resources appear to the subarea SNA network as cross-domain resources. As these resources have always been dynamic, VTAM developers used this method to allow the APPN resources to appear on a dynamic basis.

Figures 2.11 and 2.12 show the way the composite node appears at different points within the APPN network. Figure 2.11 shows the detailed view of the actual resources within the APPN network. Figure 2.12 shows the logical view of the APPN network. As the VTAM and NCP disappear into a composite resource, the details of the network disappear.[10]

---

[9]Even this was provided through the use of some advanced techniques by VTAM and NCP.

[10]This loss of network detail shows up again later in Chap. 6. This loss of detail can lead to some difficulty in diagnosing network problems.

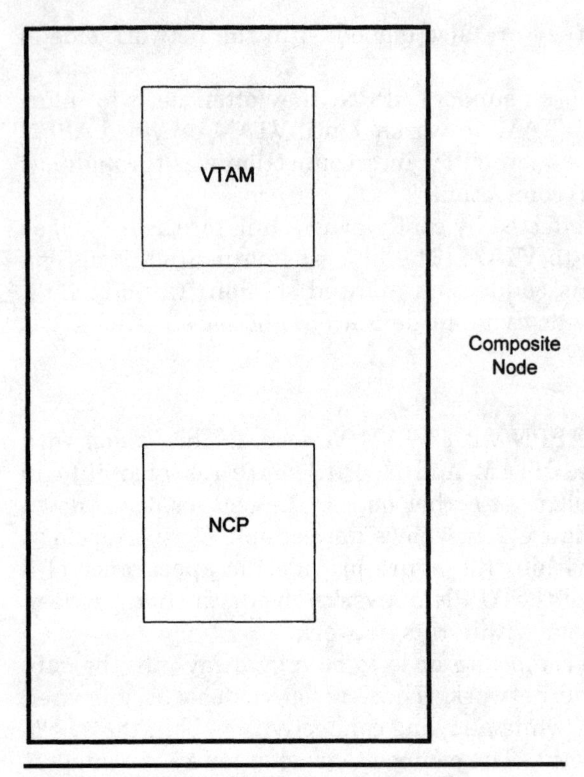

Figure 2.10 VTAM and NCP as composite node.

2.4.3 OS/2 communication manager

Operating System 2 (OS/2) Communication Manager (CM) provides APPN functions for both ENs and NNs. The ability of OS/2 CM to provide an APPN appearance allows APPN to be moved into areas that do not require an expensive computer, such as a System/390 or AS/400.

Though CM is able to provide network node functionality, it would not be advisable to use an OS/2 workstation to provide NN functions for an APPN network that contains a large number of NNs. The overhead associated with the NN features would overwhelm the workstation.

2.5 APPN networks

APPN networks can either be pure or a mixture of APPN and subarea SNA resources. Each of these types of APPN networks operate in a similar manner. Both allow the dynamic nature of APPN to gain awareness of the network and will fulfill the architecture.

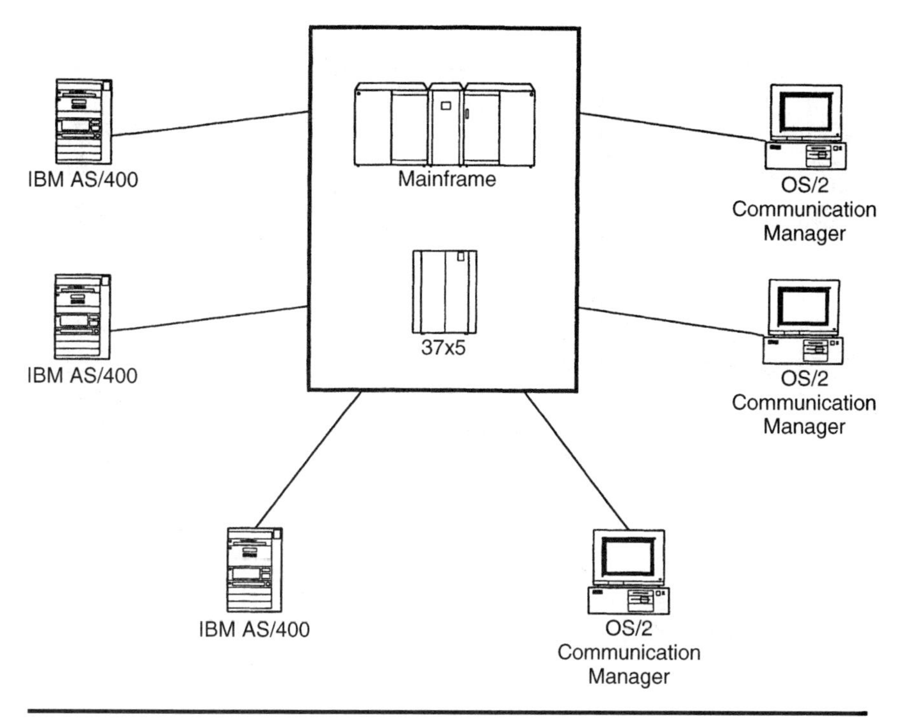

Figure 2.11 Composite node physical components.

The reason for these two types of networks is the requirement for migration from the subarea SNA architecture. As some subarea networks contain tens of thousands of resources, quick migration is not only not feasible, but is often impossible.

2.6 Summary

APPN has grown out of the demands of network users. Though the architecture has been available for quite a while, it was not until recently that APPN came into its own and IBM moved its considerable influence fully behind it.

The architecture provides many of the features that network users and managers required. These include:

- Ease of use
- Ease of management
- Flexibility in implementation

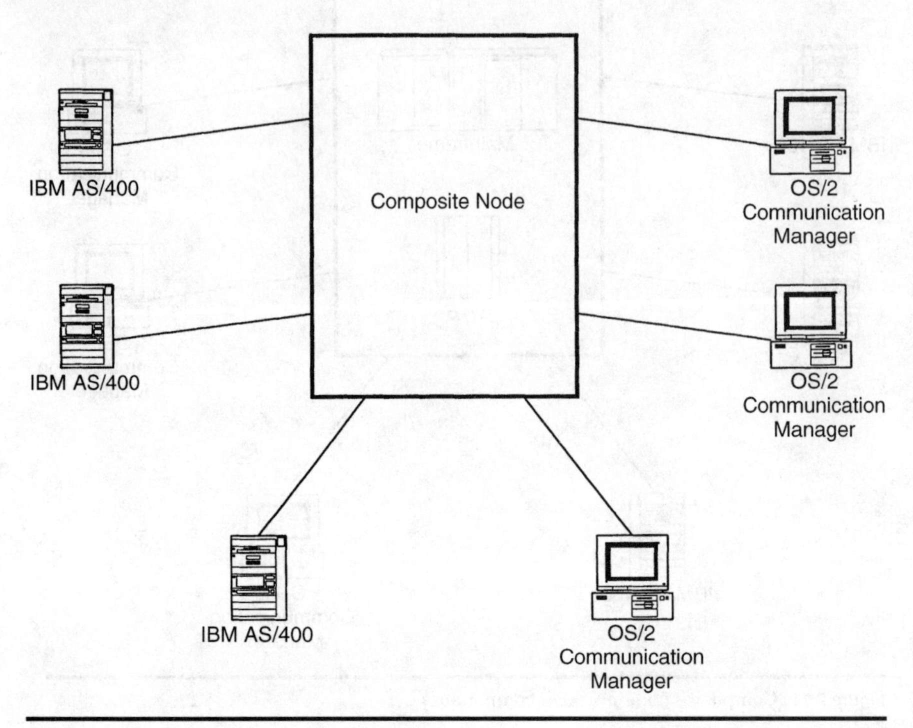

Figure 2.12 Composite node logical structure.

- Dynamic routing
- LU registration

Because it is even more fully architected than subarea SNA, APPN has been able to expand into other platforms at a quicker pace than subarea SNA did. The strong architectural basis of APPN has allowed the design to prosper and expand into new areas without resulting in incompatibilities and inconsistencies.

APPN networks consist of three basic types. These are LEN, EN, and NN. In addition, the combination of VTAM and NCP can provide the appearance of an APPN node to other nodes within an APPN network. This combined node is known as a *composite node*.

The APPN node has been designed with a strong architectural viewpoint. This has led to the formation of a node structure that provides clear interfaces. These interfaces fulfill the need for expansion of the architecture and the development of a code base that provides the functionality of APPN.

In order to provide APPN functions, several new SNA formats were required. These include an expansion of the XID passed between nodes to allow identification. The expansion of the XID into the type 3 format allows for the dynamic nature of APPN. Nodes can adjust how they appear to others within the network and how the network is structured. This can be done by changing the session requirements.

The AS/400 is where APPN functions where first realized. By using AS/400 networks as the basis for APPN, the architecture was able to grow into maturity. The architecture was then implemented by OS/2 CM and the 3174 cluster controller.

Only after most of the difficulties were worked out was APPN extended into the VTAM and NCP arena. The composite node was created by utilizing features of VTAM and NCP that allowed the migration to APPN to be orderly. It also was designed to support resources in any combination of subarea SNA and APPN nodes. In some cases, this support may be highly complex, but a good design allows almost any customer configuration to be fully supported.

Chapter

3

High-Performance Routing (HPR)

This chapter discusses the newest network architecture treated in this book, High-Performance Routing (HPR). This architecture is an extension of the base-APPN that provides some key advancements. These new functions include:

- Nondisruptive path switching
- Better utilization of high-speed communication paths
- An advanced congestion control methodology

Because HPR is an extension of the base-APPN functionality, all HPR nodes support APPN functions. As a result, HPR nodes can operate within a non-HPR network as another APPN node. The only difference is that the HPR node provides some control vectors on certain flows that non-HPR nodes would not recognize. Since unknown control vectors are ignored, this will have no effect on the operation of the APPN network.

3.1 Why HPR?

When APPN was developed, the paradigm of network professionals consisted of low-speed analog links connecting nodes within the network. These low-speed connections allowed a full protocol stack to be implemented within each node in the network.

With the advent of high-speed communication, the overhead associated with using a full-protocol stack within every node, including intermediate nodes, could not be tolerated and still be expected to utilize the full bandwidth available. What was required was a protocol stack that pushed the higher-level session control to the outer edges of the network while utilizing a smaller stack *within* the network. A merging of the predetermined routing supported within subarea SNA with the dynamic route determination provided by APPN was needed.

HPR is such a marriage of functionality. It is also an extension of the APPN architecture that allows for a reduced transport stack in routing nodes; that is, nodes that are not the termination point for a session. HPR implements a method of pushing session-level considerations to the outer edges of the network.

One of the shortcomings of APPN is that it does not support the rerouting of session data around a failure in the network. APPN requires that a session outage be recognized by the session endpoints. Only after a session has been disrupted is a new route calculated and the failed session reestablished. This disruption in the session is seen not only by the session endpoints, but by the end user. This is obviously not the best method of keeping communication available to the end user. Such a situation contrasts with the operation of an TCP/IP network.

In HPR, packets are automatically rerouted if a transport failure occurs. The end user has no awareness of a session outage because packets are automatically rerouted around a failed component by the Internet Protocol (IP) layer.[1] This is because of the dynamic hop-by-hop routing nature of IP. If a link failure occurs, that route segment is taken out of the topology and packets are routed in another direction.

3.2 Architecture

Additional functionality of HPR is provided by two new components: rapid transport protocol (RTP) and automatic network routing (ANR). These components provide the added functionality exhibited by HPR nodes.

3.2.1 Rapid transport protocol (RTP)

RTP is a full-duplex, connection-oriented protocol designed to support high-speed, low-error-rate communication paths. This is provided by

[1]If packets cannot be delivered, the Transmissional Control Protocol (TCP) session eventually fails. Such an occurrence is seen by the end user as a communication outage. Thus, Transmission Control Protocol/Internet Protocol (TCP/IP) can include session failures that are caused by a different scenario.

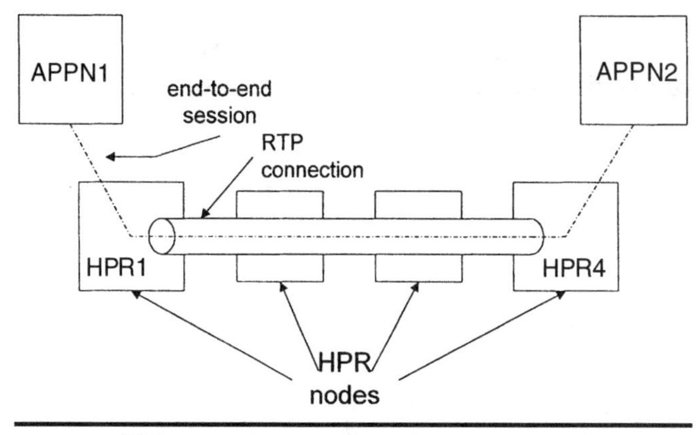

Figure 3.1 RTP connection spanning nodes.

establishing RTP connections between origin and destination points within the HPR subnetwork.

RTP connections are virtual pipes that are created between HPR endpoints. The virtual pipe provides for the nondisruptive path switch that allows sessions to bypass network outages automatically.[2]

Figure 3.1 shows an RTP connection. It forms a virtual pipeline across the network. The RTP connection is bounded by HPR3 and HPR4. Note also that the application session is between APPN1 and APPN2.

This session passes across the APPN network to the HPR subnetwork. At that point, the session flows across an RTP connection to the exit point of an HPR subnetwork, where it reenters the APPN network. This session eventually reaches its destination of APPN2.

RTP connections are tied to the endpoints of the connection and the class of service. If a connection already exists between two HPR nodes that satisfies a session request, the same RTP connection is used, rather than starting a new connection. Thus, the number of RTP connections that must be maintained are reduced.

Between APPN1 and HPR3, normal APPN flows are used. This allows APPN nodes to be supported without requiring any modification. In fact, the APPN nodes have no awareness that the HPR nodes are any different. Route calculation, directory searches, and APPN topology operate in the same manner as before HPR.

---

[2]It should be noted here that only failures along the RTP connection are rerouted. If the failure occurs between APPN and HPR nodes, the session will be lost.

NOTE: It should be understood that an RTP connection is not a session, because no BIND flows between connection endpoints. However, it is a connection-oriented protocol that assumes most of the appearances of a session. These include connection awareness, sequencing of packets, requests and acknowledgments of packets, pacing of data, and an end-to-end transport appearance.

RTP also provides for a change in the method of error recovery along the RTP connection. Instead of requiring error detection and recovery on each link along a path, RTP provides for end-to-end error recovery. The overhead associated with providing error recovery for each link is thus reduced.

The assumption is that communication error rates have become so low that the overhead associated with a retransmission across the RTP connection is less than the error recovery for each link.[3] When high-capacity links are being utilized, this is the only effective method of transmission, because the time associated with error recovery will slow throughput on the link sufficiently to eliminate any gain that the high-speed connection can provide.

The last addition with RTP is for end-to-end flow and congestion control. RTP uses a new method of pacing that predicts congestion and takes the necessary steps to decrease the probability of congestion from occurring. This type of pacing is called adaptive rate-based congestion control (ARB). By using this method of congestion control, the processing required in each node to react to congestion is reduced and higher link utilizations are tolerated.

3.2.2 Automatic network routing (ANR)

The advent of high-speed communication required a different routing algorithm that would be fast enough to provide full utilization of these connections. ANR provides this by supporting fast packet switching and not requiring any session awareness in intermediate nodes. Without these features, high-speed connections cannot be adequately utilized.

Fast packet switching, at the routing level, is a source routing protocol that is similar to the type of routing provided by token ring networks. An end-to-end path is chosen for a session by the originator of a request. There is no dynamic rerouting of frames within this con-

---

[3] This is similar to the view of frame relay proponents.

nection.[4] Instead, each intermediate node uses label routing to route between nodes.

When using label routing, a routing decision is not dynamically made. Instead, a source route is preassigned to a route. When an intermediate HPR receives a packet, it can immediately reroute the packet because it can utilize a label that has already been assigned by the node. This label is used to determine the next hop along the path. This type of routing, instead of the normal APPN ISR processing, can greatly speed routing decisions within intermediate nodes.

ANR also eliminates storage requirements on intermediate nodes of an RTP connection by keeping no awareness of the sessions that are traversing through the node. By eliminating the 200- to 300-byte session connector per session, memory requirements for intermediate nodes are greatly reduced. This storage can be put to use by the node to contain control blocks for sessions that terminate at the node, or for some other use.

As the speed of the network increases, the time associated with routing must be reduced in order to utilize the higher-speed interface. Distributed routing protocols, such as IP, experience increasing difficulties as communication speeds increase.

HPR uses a source routing protocol that greatly reduces the processing impact of routing. The discussion of label routing shows how using fixed routes can greatly reduce the speed of routing. But this type of usage can only be supported when RTP connections are utilized. In all other cases, normal ISR processing is utilized by HPR nodes.

HPR nodes include a HPR-specific header in each packet transmitted on RTP connections. This header is used to aid in label routing. Because the route through the HPR network used by an RTP connection is fixed when the RTP connection is established, this source-routed methodology fits very well.

3.3 HPR Base and Towers

Like APPN, HPR is architected with required "base" functions and optional "towers." This methodology allows a product some flexibility

---

[4]This statement may seem like a contradiction, but it is not. The route chosen for an RTP connection does not vary while communication is active. It is only when an error is detected that an RTP endpoint makes any change to the chosen route. It does this by building a new RTP connection. All nodes on the new path fix their labels so that routing can be done extremely quickly by using preassigned labels in the routing decision.

TABLE 3.1 HPR Base and Tower Structure

	Control flows over RTP connection option
Link layer error recovery option	Transport option
HPR base	
APPN end node or network node	

in operation and the level of complexity desired by the product. Table 3.1 shows the base and tower structure of HPR support.

3.3.1 HPR base functions

The base functions are the minimum processing required to provide HPR functionality. The majority of functions at this level are directed at support for ANR. These nodes can act as an intermediate routing node for an RTP connection.

The required base functions include:

- Intermediate routing of packets using ANR
- Support for *both* ANR and FID2 frames on the same link
- Using FID2 routing for CP-CP sessions and route setup requests
- Using FID2 routing for LU-LU sessions which use ISR
- No link layer error recovery for ANR frames
- A minimum packet size of 768 bytes

3.3.2 Transport option

The transport option allows an RTP connection to be established.[5] This option must be supported by both ends of the RTP connection. If this is not true, there is no advantage to the one node because no change in operation is possible.

This type of node must also provide an APPN/HPR boundary function. Thus, such a node must be capable of converting a session transmission from an RTP connection to a normal FID2 APPN transport. If this same node is also the session termination point, the destination

[5]Here we assume that there is more than one HPR node supporting the transport option.

LU must also be provided with the session data. If, on the other hand, the session continues along a normal APPN network, then a conversion is made to normal APPN ISR routing.

3.3.3 Control flows Using RTP connection option

The next transport tower uses the RTP connection for the transportation of control flows. These include CP-CP session flows and route request/response flows. Obviously, both HPR nodes must support this option prior to it being operational.

The use of the RTP connection for these control flows allows these requests to be transported with the advantages resultant from the RTP connection. This includes the elimination of costly ISR routing and the lower transport overhead associated with the RTP connection.

3.3.4 Link layer error recovery option

When high-speed links are used, the overhead associated with performing error recovery can become onerous. By eliminating this overhead, one can better utilize the bandwidth associated with the high-speed link.

This option eliminates link layer error recovery because recovery is on an end-to-end basis. This allows for a much lower code path length for the node when ANR is being utilized.

This option is associated with ANR and not with RTP. As a result, this option is available with or without the transport option.

3.4 Operation of an HPR Network

An HPR subnet (or network) provides the image of any other APPN node. Routes are calculated through an HPR subnetwork using normal APPN route calculations. The entire route is seen within the route selection control vector (RSCV). The topology of the network is not modified by the presence of HPR. All HPR nodes and links are seen within the topology database, just like base-APPN nodes and links.

What is changed is that the HPR nodes recognize their presence and can provide some additional functionality. Thus, if an RTP connection fails, the HPR endpoint calculates a new route through the HPR subnetwork, establishes a new RTP connection, if needed, and keeps the end-to-end session active. Communication continues across the HPR subnet[6] without the endpoints having an awareness that a new path is being utilized within the HPR network.

3.5 ANR Routing

As stated earlier, ANR is a source-routed protocol. What occurs during this process is that a route is calculated through the HPR network, as is done for a normal APPN network. It is the creation of the ANR routing area that differentiates HPR.

This area precedes the Request unit (RU). It consists of a set of ANR labels, with 2 bytes preceding the ANR labels. Figure 3.2 shows the format of these fields.

The frame is identified as a network layer packet by the first 3 bits of the frame. These are B'110', which distinguishes it from a FID2 frame that always starts with B'0010'.

Following the first 2 bytes is a variable number of ANR labels. Each label is variable in length, ranging from 1 to 8 bytes long. The string of labels is terminated by a termination label.

Each HPR node along the path uses the first ANR label to provide a quick method of determining the next hop along the path. Each node removes its own label from the list before transmitting it to the next node along the path. Thus, the number of ANR labels shrinks as the packet moves through the network. Figure 3.3 shows how this operates. Because the length of the ANR label list is variable, there are no explicit maximum number of nodes that can be traversed.

An ANR label only has local significance. There is no requirement that the labels be unique or consistent in length. They are not the same as TG numbers, so they do not appear in the topology database. It is important that they be meaningful to the local node. Possible uses of the label is an index value into a table identifying the next hop along the path or a control block address. Using one of these as the label allows for very fast switching with minimal overhead.

3.5.1 Network connection endpoint (NCE)

The last ANR label *before* the termination label is the network connection endpoint (NCE) for the path. This identifies the endpoint of an RTP connection. Because there are two endpoints, the pair of NCE labels identifies a specific RTP connection.

---

[6]All of this activity assumes that the transport tower is available on both HPR endpoint nodes.

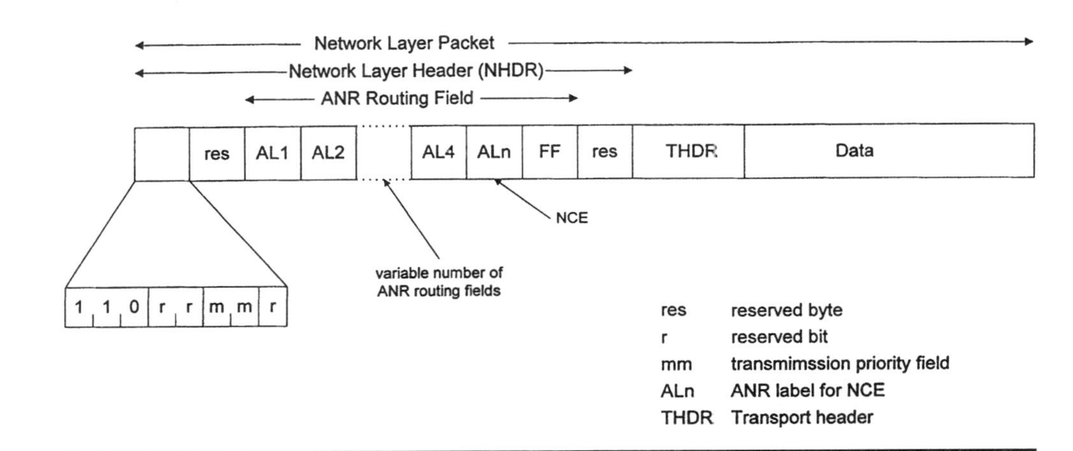

Figure 3.2 Network layer packet format.

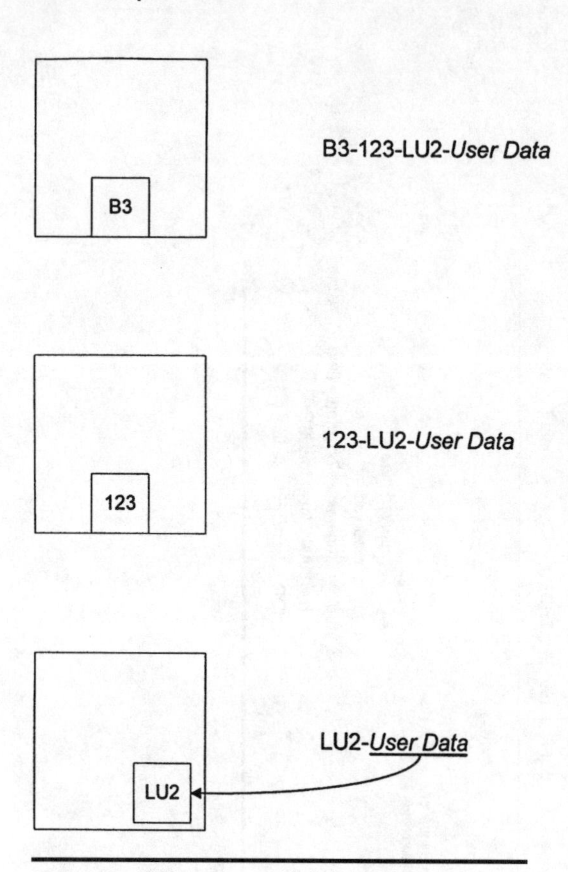

Figure 3.3 ANR routing with label reduction.

The NCE label identifies the internal component that terminates the RTP connection. This component can be any of the following:

- CP
- LU
- APPN/HPR boundary function
- Route setup function

3.5.1.1 Control point (CP). When a node implements the control flows over the RTP tower and communication is for the CP-CP sessions, the NCE label of the remote CP is used. This allows support for the internal routing of packets containing this label to terminate at the CP within the node.

3.5.1.2 LU. If the LU-LU session terminates in a node that supports the transport tower, then the NCE specifies the destination LU. When the packet arrives at the destination node, the packet is internally routed to the LU.

The label for the LU must be unique within the node. Thus, a specific LU can only have a single label. At the same time, it is possible to assign a label for a single LU, a group of LUs, or for all of the LUs within a node. This is an implementation choice affecting how specific LUs are addressed.

Although the simplest method is to have a specific label for each LU, by having a label represent a group of LUs, it is possible to use this LU group as a pool of available resources. How these LUs are allocated is an implementation choice and has no effect on the operation of the RTP connection.

3.5.1.3 APPN/HPR boundary function. If a session proceeds beyond an HPR subnet, then the APPN/HPR boundary function must be the termination point for the RTP connection. In this case, the label specifies this component within the node as the destination of the RTP connection. The boundary component can be addressed on the basis of a TG, a group of TGs, or all TGs within the node.

This component takes the HPR packet and places the required FID2 header on before transmitting the packet to the APPN network. Because the APPN/HPR boundary function resides at the end of an RTP connection, only a short ANR header is left prefacing the packet. This eases the task of conversion for transmission to the APPN network.

3.5.1.4 Route setup function. The creation of an HPR connection is provided by the route setup function within an HPR node. This protocol is utilized in order to determine if the HPR connections can be utilized and which RTP connection to be used.

Within each HPR node, the protocol is provided by packets passed between route setup functions. These packets do not proceed on sessions. Instead, an HPR node addresses the request to the route setup function in the adjacent node. This ANR label is obtained during the XID3 exchange between HPR nodes. This addressing allows a complete RTP connection to be established between HPR nodes that support the transport tower.

Communication between the route setup functions along a chosen path between NCEs provides for the definition of the ANR labels used along an RTP connection. The route setup function in each node along the path chooses a label and places it into the set of ANR labels.

The choice of where the RTP connection spans is done by the originating HPR node scanning the RSCV for the BIND that initiates the request for an RTP connection. The originating node looks to find the farthest node within the RSCV that supports the HPR transport tower while still ensuring that all intermediate nodes along the path are HPR nodes. An RTP connection will be established to the node that satisfies both of these tests.

3.5.2 Connection network (CN)

Connection networks (CN) exist and can be used within HPR networks as they do within APPN. In most ways, a CN operates the same way as for base-APPN. At the same time, some small differences exist.

Among the differences is the manner in which a CN is interfaced when using RTP connections within a CN environment. When RTP connections are used, a link must be activated between adjacent nodes so that ANR labels are assigned to the connection. Without this activation, ANR labels are not created and ANR routing cannot be utilized.

At the entrance to the CN, the local HPR node activates a TG to the exit node from the CN. This activation results in the entrance and exit nodes that generate ANR labels for their identification of the virtual TG. These labels are added to the list until the destination node of the RTP connection is reached. At that point, the route setup request is returned, along with the labels within each node. Once the RTP connection is activated, the session request is passed to the HPR/APPN boundary function NCE of the last HPR node, which routes the session request to the destination of the session request.

3.6 Transmission Priority

Not only do HPR nodes participate in normal APPN services, such as topology and directory, but they also can provide complete consistency with APPN COS processing. This is because HPR supports and uses the same topology database and directory services of any other APPN node. HPR nodes also use the same four transmission priorities as base-APPN nodes. These priorities are network, high, medium, and low.

HPR nodes create these four transmission queues for each link. If a BIND specifies a certain transmission priority, the HPR nodes along the path can support the specification within the BIND and participate in the prioritization of traffic.

Implementation-dependent aging algorithms are normally utilized within a node to ensure that low-priority traffic is not blocked from transmission. Although this is implementation-dependent, APPN nodes should provide this support. If they do not, it is possible for sessions to be lost from the lack of data flow.

3.7 Network Layer Packet

A packet transmitted along an RTP connection has a specific format. This format is shown in Fig. 3.3. It consists of three components. These are:

- *Network layer header (NHDR).* The NHDR provides addressing for the packet as it traverses the HPR network. This header contains the ANR labels used for switching packets.
- *RTP transport header (THDR).* The THDR is used by the RTP endpoints to provide correct processing of the packet. Intermediate switching nodes do not interrogate or modify this header.
- *Data.* The user data is transported through an RTP connection.

3.7.1 Network layer header (NHDR)

This header begins the frame used by HPR nodes. The components of this header include the transmission priority and the ANR labels, described previously. It is this header that is used by the ANR component of an HPR node to support the label routing utilized by HPR nodes for transport.

The NHDR consists of some indicators that identify the packet as a network layer packet. It also contains the priority field and the set of ANR labels.

3.7.2 RTP transport header (THDR)

The RTP transport header (THDR) is used by the pair of RTP endpoints that define an RTP connection. It is used for communication between the endpoints and to identify the RTP connection. This header consists of the fields shown in Table 3.2.

Each NCE generates an identifier for the RTP connection. This field, transport connection identifier, allows the HPR component to identify the RTP connection.

The THDR also contains fields that allow the NCEs to request acknowledgment of packets. This is very similar to the RH field within normal APPN and subarea SNA flows, except that the acknowledgment request and segmentation indicators are used between NCEs and not the session endpoints. These fields are used to

TABLE 3.2 RTP Transport Header (THDR)

Byte	Bit	Content
0–7		transport connection identifier (TCID)
	0	TCID assignor indicator 0 receiver assign 1 sender assign
	1	reserved
	2–63	connection identifier
8	0	reserved
	1	connection setup indicator
	2	start of message indicator
	3	end of message indicator
	4	status request indicator
	5	respond ASAP indicator
	6	retry indicator
	7	reserved
9	0	last message indicator
	1–2	reserved
	3–4	connection qualifier field indicator
	5	optional segments present
	6–7	reserved
10–11		data offset/4 position of data relative to start of THDR
12–15		data length field
16–19		byte sequence number
20–k		connection qualifier/source identifier field
k + 1 − m		optional segment field The optional segments include:
		x'0E' status segment
		x'0D' connection setup segment
		x'0F' client "out of band" bits segment
		x'10' connection identifier exchange segment
		x'12' connection fault segment
		x'14' switching information segment
		x'22' adaptive rate based segment

ensure that the communication between the NCEs is consistent and reliable.

The THDR can also contain some optional fields. These are special segments that are used for various control purposes. They include segments used during connection startup and termination, and to communicate error conditions.

3.8 Enhanced Session Addressing Using FID5

Sessions flow through an RTP connection between LUs. It is conceivable that an RTP connection could support the transportation of thousands of sessions. Although these sessions do not terminate with the RTP connection, the sessions must be accounted for in order to support nondisruptive path switching.

If the RTP connection that was used by a session fails, the RTP connection must be reestablished using the original COS. If a new RTP connection cannot be established with the original COS, then there is no route available and the session must be allowed to fail. The control blocks used for the RTP connection must be destroyed.

The FID5 address field enhances the ability to support this task. The FID5 is passed within the data portion of the network layer packet.

The format of the FID5 is shown in Table 3.3. The FID5 address field is 62 bits long. Each direction of a session is assigned a different address by the sending LU for that direction. This address is saved and utilized to address packets for a specific destination. In an effort to speed packet processing, an LU uses a control block address for identification so that the information for the session can be obtained as quickly as possible.

When the session is started,[7] the connection endpoint on the originating-side of the session assigns a session address. When this is placed into the FID5, the high bit is set to signify to the receiving side that the address is not assigned locally. This address is saved by the receiving side and another address is generated and placed into the BIND response. Now each RTP endpoint knows the pair of addresses for this session.

---

[7]That is, when a BIND is sent.

TABLE 3.3 FID5 TH

Byte	Bit	Content
0	0–3	FID5 0101
	4–5	MPF—Mapping field 10 first segment of BIU 00 middle segment of BIU 01 last segment of BIU 11 only segment of BIU
	6	reserved
	7	EFI—Expedited Flow Indicator 0 normal flow 1 expedited flow
1		reserved
2–3		SNF—Sequence Number Field
4–11		SA—Session Address (replaced OAF′, DAF′, and ODAI fields There are two addresses associated with each session; one for each direction
	0	Session address assignor indicator 0 receive assigns address 1 sender assigns address
	1	reserved
	2–63	session address session address formed by the concatenation of the node addresses

3.9 Nondisruptive Path Switch

The ability to provide a nondisruptive path switch is one of the most desirable abilities provided by HPR. This allows the path of a session flowing through an RTP connection to be rerouted in the event of a communication failure. It should be noted here that the nondisruptive path switch is only available along the RTP connection. If a failure occurs outside of an RTP connection, even at an HPR node, there will be a session outage. This limits the availability of this function.

If a failure occurs along an RTP connection and another route is available that meets the original COS requirements of the session,[8] an RTP endpoint can institute a session switch. A path switch is illustrated in Fig. 3.4, which shows the normal RTP connection path, and Fig. 3.5, which shows the same configuration after a path switch.

[8]The node must also specify in the connection setup and reply that they are willing to participate in this function.

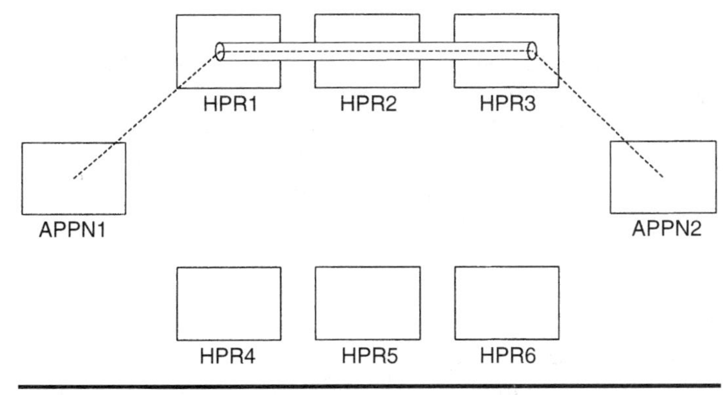

Figure 3.4 Normal HPR connection flow.

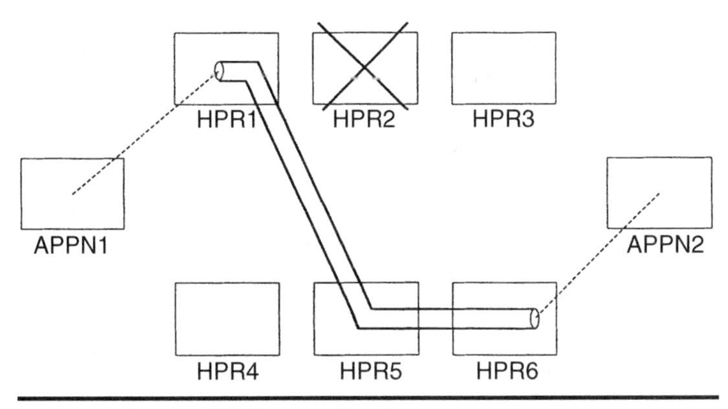

Figure 3.5 Rerouted HPR connection flow.

3.9.1 When is path switch needed?

The need for a path switch is recognized by one of several methods. The first is that the node detects an link failure. This could either be a local link failure or a topology update locates a failed remote link used by an RTP connection. Upon recognizing the failure, the node cross-references the failure to a connection, such as sessions or RTP connections, that were using the failed segment.

Another method of detecting a failure is by noticing a connection timeout. This failure is triggered by the expiration of the RTP connection alive timer. This timer ensures that an RTP connection is active. If no traffic uses a connection for a specified period of time, an RTP endpoint sends a status request message. If the response is not received, the connection is assumed to be disconnected.

3.9.2 Obtain new path

Assuming that both endpoints specify support for path switching, the first step for recovery is to obtain a new path to the NCE in the remote node. In this case, a change to the normal APPN topology search logic is required so that the chosen path uses only HPR-capable links, even if a better path is available using non-HPR connections.

If another RTP connection is available that meets the COS requirements of a session, it will be utilized. If no existing RTP connection meets these requirements, a new RTP connection will be established.

Assuming that a new RTP connection must be created, the connection must be established by passing a request along the newly calculated route, but using ANR labels of the adjacent route setup function. Each route setup function along the path participates in the creation of the new RTP connection.

If the route setup is completed in time (see Sec. 3.9.3), a new RTP connection is established. Sessions are assigned to appropriate paths so that endpoints have no awareness of the path switch within the HPR network.

A limited resource path is used within an APPN network as a method of utilizing a facility, such as a switched facility, to overcome an immediate requirement. When this type of facility is used, each session endpoint monitors the limited resource to determine its usage count. If it is defined as such, when the usage count drops to zero, the session endpoint disables the path, by such action as hanging up the switched facility.

If a limited resource path is used for an RTP connection, the session endpoints are not aware of the path switch, so they cannot monitor the usage count on the limited resource path. This results in the possibility of a limited resource connection not being automatically deactivated when its usage count reaches zero. Although this does not have a large effect, this monitoring of the limited resource path is more difficult. It also makes the administration of this type of path more problematic.

3.9.3 Path switch timer

When a path switch is being performed, a path switch timer is started. This timer specifies a limitation on the time span allowed for the path switch to be completed. This prevents a path switch from being unsuccessfully retried forever.

3.9.4 Path switch limitations

The most important issue to be remembered is that the path switch option only applies to the RTP section of a route and is only attempted if both the endpoint nodes specify support. If a failure occurs elsewhere along a path, this function will not provide any protection. Dependent on your configuration, this may or may not be of concern.

If your network consists of either base-APPN nodes, with a small number of HPR nodes, has HPR nodes that do not support the HPR transport tower, or contains nodes that do not support path switching, the configurations in which you will be provided this feature will be limited.

There can also be a problem if you are using an EN that contains the HPR transport tower, but that has a network node servicer (NNS) that is a base-APPN node. If the original route uses only HPR nodes, an RTP connection can be utilized. If a link failure occurs, the HPR-EN requests a new route to be calculated, but the base-APPN NNS does not recognize the specification for HPR-only nodes, so it generates a route that does not meet the HPR requirement. In this case, the EN must verify the RSCV to ensure that it contains only HPR nodes.

A way to stop such a situation from occurring is to ensure that the NNS for all HPR-ENs is also an HPR node. In this case, the NNS would recognize the "HPR-only" specification and would generate an HPR-only route.

3.10 Adaptive Rate-Based Congestion Control

ARB congestion control provides control from congestion by controlling the speed data enters the network. In addition, this algorithm attempts to predict congestion, rather than reacting to it after congestion has occurred. This provides better management of the network than reactive congestion control can provide. The objective of ARB is to regulate the network traffic to a "smooth" traffic pattern, instead of having sharp peaks and valleys in network throughput. By providing a smoothed pattern, throughput becomes more predictive and the user perceives the network as more stable.

ARB determines throughput rate differentials at the sending and receiving side of an RTP connection. Because RTP connections are full-duplex, each end of the connection operates as both a sender and receiver to the network.

As an RTP endpoint starts to detect rising congestion, it slows the transmission rate. Feedback to the other RTP endpoints is provided through both the use of ARB optional segments within the THDR and its own monitoring of the network traffic patterns.

3.11 APPN/HPR Boundary Function

In an effort to provide a migration path and method for integration with existing APPN networks, HPR allows for an APPN/HPR boundary function. This enables an HPR node to provide a point at which packets can enter and leave the HPR subnet.

Because base-APPN networks use FID2 frames with ISR routing and HPR networks use an optional RTP connection using ANR for level 2 switching, there must be a translation point within the network. This translation point is the APPN/HPR boundary function. It is here that the translation is made between the APPN and HPR subnets within the larger network.

3.12 HPR Limitations

HPR nodes provide a great deal of enhanced connectivity, but they still possess limitations in their ability to support various configurations and migration scenarios. Among the limitations are:

- Path switching is limited to the segment of the path using an RTP connection
- Lack of MLTG

3.12.1 Path switching limitation

HPR comes with the promise of rerouting traffic around a failed component in the network. This promise is only partially fulfilled because the path switch function is only applicable to a potential small portion of a route. Because this feature is only available along an RTP connection, all base-APPN nodes gain no flexibility. In addition, because the RTP connection is predicated upon a minimum of two HPR nodes supporting the transport option, not even the use of HPR-capable nodes guarantees path switching.

As a result of this, your particular configuration will dictate the gain realized when using path switching. It is possible to have a large network that contains a high number of HPR nodes, that have little or no gain from this feature.

The gain is dependent on whether sessions are using RTP connections and the placement of LUs on the network. If the LUs are on nodes that are the NCEs of RTP connections, then the gain can be large.[9] On the other hand, if the LUs are on nodes that require the use of ISR connections to the end of RTP connections, it is possible to gain little, especially if the failure occurs outside of the RTP connection.

If a path switch occurs, the session endpoints have no implicit awareness. As a result, it may not be possible to automatically detect the use of a switch facility. It is also possible that the switched facility will not terminate when the usage count decreases to zero, as would normally occur.

3.13 Summary

HPR is an extension of base APPN functionality. It is extended through the use of two new components. These are RTP and ANR.

HPR is actually more than a single architecture. Instead, it is a set of architectures that provide additional functionality that is supported by base APPN nodes. The architectures are provided in a typical base-and-tower structure. The towers defined by HPR include:

- *Transport option.* The transport option allows the creation of an RTP connection between HPR nodes.
- *Control flows using RTP connection.* This option allows for typical APPN control flows to flow across RTP connections. If this tower is not supported, these control flows use a FID2 transport. This tower requires that the node also support for the transport tower.
- *Link layer error recovery.* This tower reduces error recovery along each connection. By reducing the overhead associated with error recovery at each node, higher speed connections can be supported. This tower is available with or without the transport tower.

RTP provides a full-duplex connection between endpoints within an HPR subnetwork. This connection provides a virtual, end-to-end pipe through which session traffic flows. This is known as an *RTP connection.* This pipe also allows for three major additional functions for these nodes. These functions are:

---

[9]We are assuming that intermediate nodes are HPR-capable.

- Nondisruptive path switch
- End-to-end error recovery
- End-to-end flow and congestion control

A nondisruptive path switch allows the path of the RTP connection to be modified without notifying session endpoints that are using the connection. Thus, the path of the RTP connection can be changed, and with it the sessions that are flowing through it, without disrupting the sessions that are utilizing the connection. Thereby, the session path can be changed to bypass a network outage without disrupting session flow. This constitutes a major advancement for these nodes. The only problem with the above function is that it is only applicable to the portion of the session path that traverses along an RTP connection, which only flows between HPR nodes that support the transport option. The session path that falls outside of these nodes does not gain the same functionality.

End-to-end error recovery provides for the elimination of error recovery at each intermediate node along a session path. Instead, error recovery responsibility is limited to the endpoints of an RTP connection. A reduction in processing overhead at the intermediate nodes is thus allowed. This results in the ability to better utilize higher-speed connections than was feasible in base-APPN ISR nodes.

Although the above procedure requires error packets to be retransmitted across the entire RTP connection, two factors reduce the impact of these retransmissions. The first is that very high-speed digital links have very high reliability profiles. The probability of needing error recovery caused by a transmission error is thus reduced. The second factor is that the RTP connection supports selective retransmission. The result is retransmission of only the packets in error, not an entire sequence of packets.

RTP is a tower on the base functionality of an HPR node. In order to support RTP connections, the node must support the HPR transport optional tower. This is a tower over-and-above base HPR functionality.

ANR provides for the new routing mechanism for HPR nodes. The purpose of ANR is to reduce the overhead associated with intermediate routing nodes within an APPN network. The reduced overhead is a result of three key ANR functions. These functions are:

- Source routing
- Fast packet switching
- No session awareness in intermediate routing nodes

Routing through the HPR subnetwork using an RTP connection uses a source routing algorithm. Because the path of a route is preestablished, there is no overhead associated with determining the route of a packet through the HPR subnetwork.

The source routing that is used creates a set of labels that define the route of the RTP connection. These labels are used by each HPR node to pick the next segment along a route. It is recommended that the label used, which is a variable-length field of 1 to 8 bytes picked by the intermediate node itself, be chosen intelligently so that switching can be done as quickly as possible. One possible choice for a label is a control block address that defines the next segment along the path. As the node provides the switching, it also eliminates its own label from the list of labels used to provide intermediate switching. The speed of switching is thereby increased by requiring each node to process only the first label in the list. Thus, there is no requirement to scan the list in an attempt to locate a node's own label.

The intermediate switching nodes have no session awareness when using ANR routing. In this configuration, only the endpoints of an RTP connection have this knowledge. This eliminates the need for a session connector control block, which occupies 200 to 300 bytes per session, for each session in each routing node. Such a requirement can overload the storage capacity of some nodes. Such a situation is especially critical because switching nodes will start to support hundreds of thousands or millions of sessions.

HPR nodes support the use of new packet formats from those supported by base APPN nodes. These formats include a network layer packet, which includes a network layer and RTP transport header, and a FID5, which can be included within the user data portion of a network layer packet.

The network layer header contains the fields used by an HPR node to support the label switching described for HPR nodes. This header includes the labels that must utilized by HPR nodes in order to provide the lower-level switching allowed by HPR nodes.

The RTP transport header contains fields that allow communication between RTP connection endpoints. Included are some control vectors associated with RTP endpoints, and fields that allow endpoints to request information from its counterpart on the other end of the connection.

The FID5 allows support for the lack of session connectors within HPR nodes.[10] Instead, the FID5 uses an enhanced session addressing

---

[10]HPR nodes support normal APPN functionality with ISR routing, but I am speaking here of additional capabilities supported by HPR nodes.

method that allows an RTP endpoint to support thousands of sessions using the connection without exceeding memory capacity.[11]

Nondisruptive path switching is a key added function provided by HPR nodes.[12] This feature supports path switching of sessions using an RTP connection. The main limitation is that the path switching only applies to the portion of the session flowing along the RTP connection. If an outage occurs outside of the RTP connection portion of the path, nondisruptive path switching will not be allowed.

In the case of a path switch, one of the RTP connection endpoints recognizes an outage along a connection path. Because it has some knowledge of the sessions flowing through the connection, the node is able to generate a new path for each session. These paths use the original COS information, but it is specified that only HPR nodes can be utilized along the new path. If such a path can be created, the path of the session will be switched, but no indication will be passed back to the session endpoints that the path switch has occurred.[13]

---

[11]Although the RTP endpoint does not have a complete session connector, some session awareness is needed in order to support nondisruptive path switching. Without some session awareness, the endpoint would not be able to determine which sessions to switch.

[12]This feature requires support for the transport option.

[13]The new path will be unknown to the session endpoints. Each endpoint will only have awareness of the original RSCV that was created for the session.

Chapter

4

Internet Protocol (IP)

Internet protocol (IP) has been gaining widespread acceptance over the last 5 years. This protocol has been a driving force behind the expansion of communications throughout the world.

4.1 History

Internet Protocol traces its roots back to the late 1960s. At the time, the Advanced Research Projects Agency (ARPA) of the U.S. Department of Defense (DOD) began the development of the communications system that was the forerunner of today's Internet. ARPA, in partnership with researchers and universities across the United States, developed new communications technologies.

Advanced Research Projects Agency (ARPA) and its network, ARPANET, had a different set of requirements than most users of communication systems at the time. Among the requirements were:

- *Survivability.* ARPANET was an outgrowth of the DOD. As such, the method of communications was directed at its ability to survive disasters (such as war). As this was being done during the height of the Cold War, thoughts of surviving various levels of armed combat was paramount during the development process of protocols.
- *Dynamic.* The survivability of the communications system was seen as relying on the ability of the system to be dynamic. If a section of the communications structure was not available (e.g., a set of communication lines was not active), the protocols should be able to dynamically adjust and discover a path between the endpoints of the system.

- *Distributed.* In an effort to enhance the survivability of the communication infrastructure, any single point within the network should not cause the unusability of the rest of the system. Not only were the underlying protocols distributed, but components within the network should also be distributed in nature.
- *Support for the interconnection of resources.* Separate networks were created across the United States. Because each of these networks were isolated, there was a desire for interconnection between the networks to be recognized. A method of enabling communication across network boundaries was necessary while keeping management of each network independent. Yet, researchers wanted freedom to pass information easily across those network boundaries.
- *Independence from hardware or software.* Many proprietary communication protocols had been developed. Each protocol was dependent on a specific set of hardware and/or software. These proprietary methods resulted in limitations on how the systems could be joined, as well as the inability of some systems to communicate at all! In an effort to remove these artificial restrictions, proprietary protocols were removed from consideration in designing the protocols for the new network.

The result of this effort was ARPANET which, by the early 1980s, consisted of over 300 computers. These systems were largely able to intercommunicate and were free to support and develop new protocols.

By 1990, the original ARPANET was absorbed into what is now called the Internet (capital *I*). This is a loose collection of computers that numbers in the thousands but supports communication between millions of users throughout the world.

The remarkable characteristic of the Internet is that it is largely an informal collection of systems that continues to be updated on an hourly basis. Computers are added, removed, or moved in their connectivity. Yet, thousands rely upon the Internet to provide communication throughout the world. Among the users are researchers, university personnel, and school children. Together, they are able to create a network that is so large and diverse that no one is able to fully understand what is available.

4.1.1 Protocol acceptance

In 1982, the DOD decided to standardize the protocols that underlaid the ARPANET. They also created the Defense Data Network (DDN) to oversee further development of the network. In 1983, the DOD modified its direction by standardizing on Transmission Control Protocol/Internet Protocol (TCP/IP).

TCP/IP—riding upon the infrastructure that was created by the DDN, other government agencies, and departments—accepted the DDN protocols. With this large base of users and the intermingling of commercial users that were contracted by the U.S. government, the TCP/IP architecture spread throughout many companies and users. In addition, by having the U.S. government demanding TCP/IP usage in their computer procurements, the demand for this technology was a large enticement.

The DOD also aided to the availability of TCP/IP by funding Bolt, Beranek, and Newman (BBN) to create TCP/IP for UNIX.[1] This code was used by the University of California at Berkeley to add TCP/IP into their variation of UNIX, Berkeley Software Distribution (BSD) 4.2. These dialects of UNIX were recently merged into a common base called AT&T's System V.

4.1.2 Method of protocol creation and acceptance

The protocols used within the Internet are coordinated by an independent organization called the Internet Architecture Board (IAB). The IAB identifies technical areas that require further research and development. The IAB also organizes the diverse activities within the Internet to create new and enhanced protocols for the Internet itself.

The IAB manages two subgroups. These are the Internet Research Task Force (IRTF) and the Internet Engineering Task Force (IETF). The IRTF coordinates long-term projects, while the IETF provides leadership on more immediate requirements. Both the IRTF and IETF create working groups that do the actual work.

These working groups consist of volunteers that have the expertise to tackle a specific project. The working groups use a method that combines both theory and practical implementations. By using an iterative process of design–develop–test–review, ideas are constantly being tested and refined against real problems. This method enables flaws and oversights in the design to be constantly tested. Since testing is done against actual implementations, the code becomes refined. Because many of the implementations become public domain, the

---

[1]UNIX was created by AT&T researchers. Though AT&T owned the code, it was made available for free to researchers throughout the country. The University of California at Berkeley modified the AT&T code to create a dialect that was widely accepted.

acceptance of the protocol is accelerated by enabling other working groups to implement and enhance their protocol easier. This speeds the development cycle, encourages the spread of the protocol, and aids interoperability.

Both university research groups and commercial organizations have also made significant contributions. These ready-made solutions are accepted by working groups members if the solutions are useful.

4.1.3 Request for comment

The Internet community is open to new ideas. This is how it has been able to develop so quickly and uses its influence within the Internet community. An example of its openness is seen in its ability to develop and propagate many new ideas.

One of the ways that this is done is through a request for comment (RFC). An RFC is a document that describes a particular feature that has been developed. The purpose of this document is exactly as the name implies; it is meant to be commented on by other users. Through the iterative process of idea development, creation of an RFC, commentary, and new development, ideas are brought forth and refined. The fact that these ideas are the product of potentially thousands of users working on its development results in the development of good ideas that are both usable and useful.

There are two basic types of RFCs: informational and standards. Informational RFCs are meant to be an example of how a technology can be developed. The purpose is to describe a methodology of development. Normally, these are close-ended in that they are not meant to be the basis for further development.[2]

Standard RFCs are meant to fill a niche and are done in the iterative process. These RFCs are distributed for the purpose of allowing others to make comment and refine the ideas within the RFC. The end result of this process is a new standard that is used by everyone.

4.2 Architecture

TCP/IP is similar to the previous network architectures discussed in Chapters 1 and 2. TCP/IP has not defined all of the OSI layers, like SNA and APPN, but much of the OSI functionality is included in the

[2]This is not be say that informational RFCs are not open to further development. It is just that this is not their intended purpose.

full stack. In addition, TCP/IP has purposely been kept "thin" to allow more flexibility in its structure and implementation.

4.2.1 Layers

Figure 4.1 provides a high-level view of where TCP/IP falls within the OSI seven layer model. This figure shows that the delineation between layers is not exactly the same in OSI and TCP/IP. Specifically, TCP/IP crosses into the Session layer, as described within the OSI model.

TCP/IP is based on the packetized routing of network data. Data is packetized, if necessary, into smaller units. These common length packets are routed through the network toward the destination.

The use of a layered architecture, as was true of subarea SNA and APPN, allows greater flexibility in the development of new interfaces and facilitates development of the protocol.

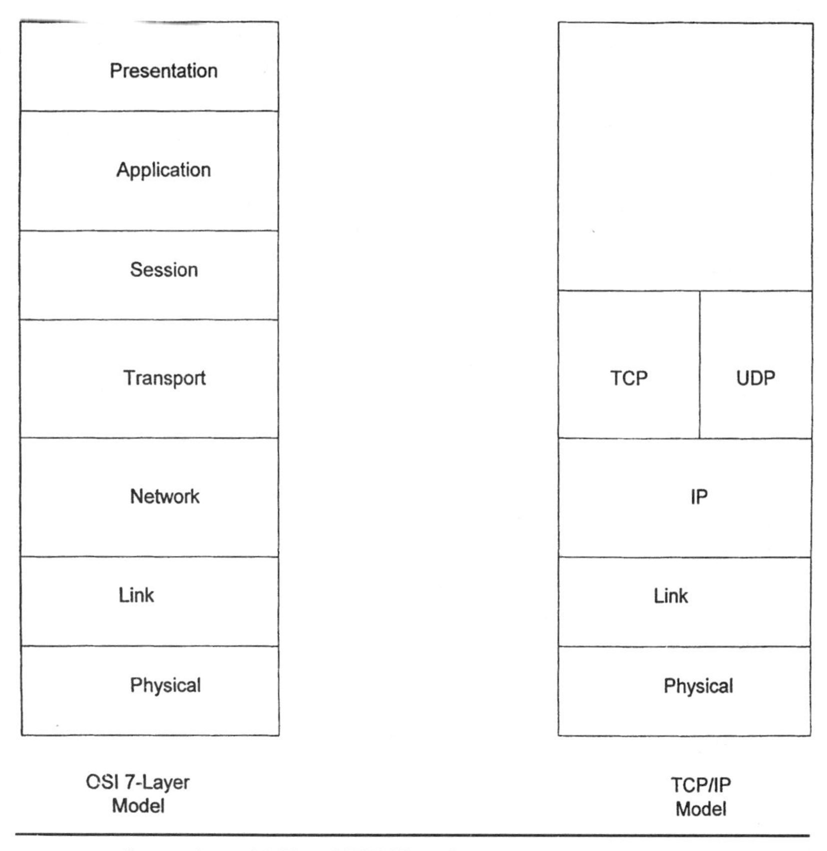

Figure 4.1 Comparison of OSI and TCP/IP architecture.

4.2.2 Protocol overview

TCP/IP can be broken down into four layers. These are the physical and link layers (combined into one layer), the IP or routing layer, the TCP/UDP (user datagram protocol) layer, and the application layer. By defining the interfaces between each of these, TCP/IP has been able to support a diverse assortment of computer systems and link types.

4.2.2.1 Physical and link layers. This layer of TCP/IP consists of the physical interfaces, such as ethernet network interface cards (NICs), RS-232, serial connections. These define how the electrical interface is provided and how data is encoded upon the media. In addition, this layer describes how data is passed onto the media and how data is transported across a single physical link.

Data at this layer is encoded onto the media, such as unshielded twisted pair wiring or 10BaseT. It is also passed from the local interface to another interface across a single media connection. Thus, data is not routed at this layer, but only received at a remote connection to the same physical media as the local interface.

Packetization is done at this layer. The actual message may be larger than the maximum size allowed for the physical media. If this is true, this layer splits the message into pieces that are passed to the next higher layer, IP, for routing to the destination.

4.2.2.2 Internet protocol (IP). Internet protocol (IP) is used to define the routing of traffic from an origin address to the destination address. Internet protocol routes each packet to the destination. As IP is connectionless, the packets are routed independently of each other. There is no requirement that component pieces of a message must travel along the same path.[3]

IP also does *not* guarantee that packets will arrive at the destination in the same order, or that they will arrive *at all*. There is no delivery confirmation or guarantee. The purpose of the IP layer is to attempt to route the message to its destination.

Figure 4.2 shows a set a packets and how they are routed through the network between Node A and Node B. Internet protocol determines the route of each packet independently for each of the packets. For this reason, plus the fact that IP makes no guarantee for the order of packet delivery, the packets arrive at Node B in a different order than they left Node A.

---

[3]This was done as a result of the survivability requirement of the ARPANET.

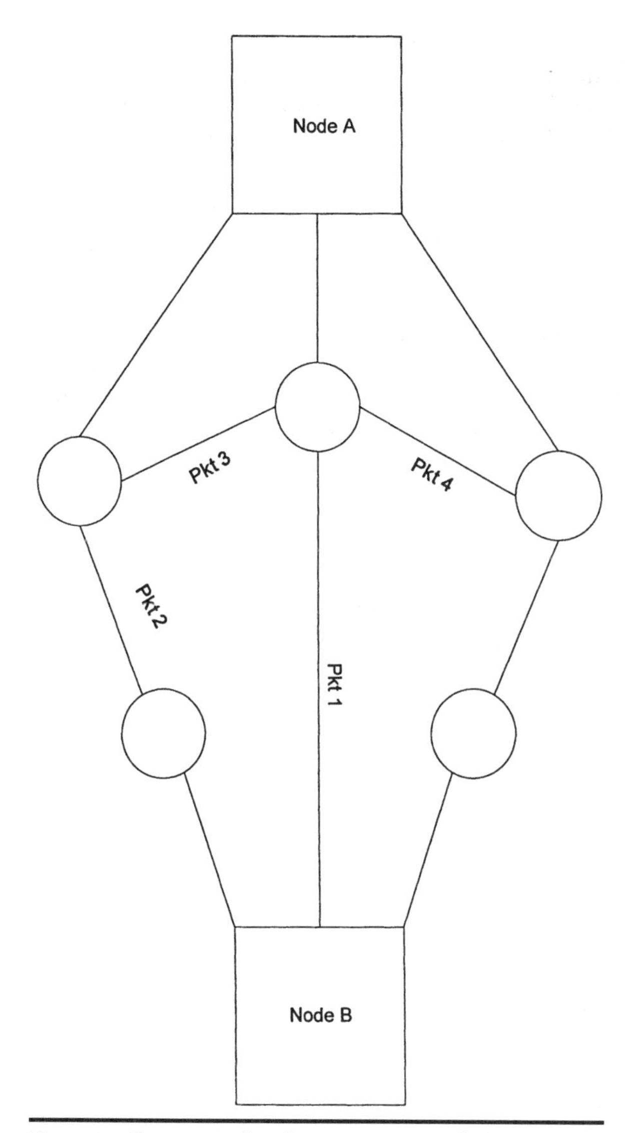

Figure 4.2 Transport of packets across network.

4.2.2.3 User datagram protocol (UDP). User datagram protocol (UDP) is one of the transport protocols used on top of IP. UDP provides a connectionless transport protocol. Packets sent by the higher application levels pass messages to UDP to be transported. As UDP provides no packet delivery acknowledgment or packet resequencing, there is no guarantee that packets will arrive at the destination nor that they will arrive in the proper order.

With these characteristics, UDP should not be used to create a reliable connection. Instead, UDP is used when the arrival of a packet is not seen as critical and where the lower overhead of UDP (in comparison to TCP which will be discussed next) is advantageous.

4.2.2.4 TCP. TCP is a reliable transport layer used on top of IP.[4] TCP provides a session between communication endpoints. This session allows for packet reassembly for higher layers and assures that messages arrive at the destination.

TCP is the most prevalent transport protocol. It is used by most higher-level applications, such as File Transport Protocol (FTP). This is because of the reliable transport provided by TCP. Such items as message delivery, packet reassembly, and message retransmission are provided by TCP. The higher level applications see only a reliable transport that ensure that messages arrive at the intended destination.

4.2.2.5 Internet control message protocol (ICMP). Internet control message protocol (ICMP) messages are used between IP nodes. Support for the reception and transmission of ICMP frames is required for all IP nodes. These frames are carried within an IP envelope.

ICMP is used for the transmission of some IP error messages and to pass some control messages. The most prevalent ICMP is the ping request and response.

ping is a small error determination tool that is used to determine if connectivity exists to a specific destination. By pinging to different nodes along a route, it is possible to determine *where* a discontinuity exists on the path. This is shown in Fig. 4.3. By pinging to nodes along the route, it is possible to see that Node D was either not receiving the ping request or the response was not being returned to Node A.

4.2.2.6 Address resolution protocol (ARP). When IP nodes communicate, they must first determine where the destination node resides and how to route messages to that destination. Unlike subarea SNA and APPN, IP routing is completely dynamic. In this mode, there are no predefined paths between nodes. Instead, the IP nodes must determine how to reach the specified destination. One of the methods used is to use address resolution protocol (ARP) frames.

When a request is sent toward a destination, the IP layer determines if it already knows how to reach the specified destination. If it does not, an ARP frame is broadcast to all nodes. This broadcast

---

[4] Thus comes the name *TCP/IP*.

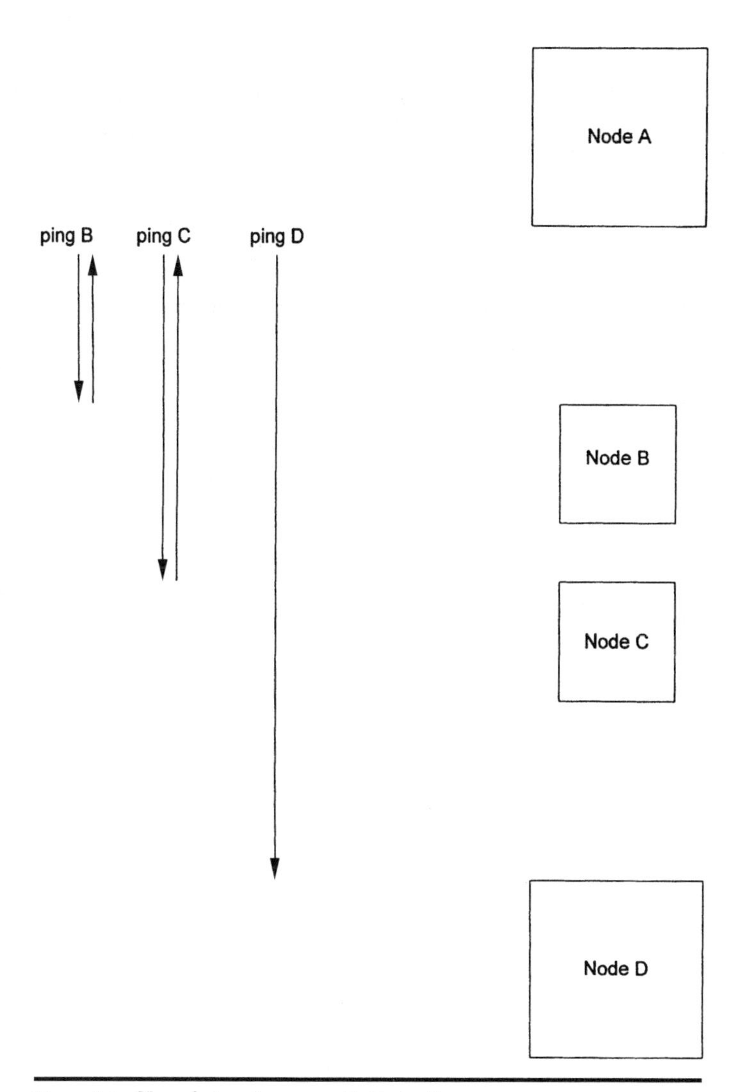

Figure 4.3 Use of ping for connectivity testing.

frame is propagated by all nodes until one knows how to reach that destination. At that point, an ARP response is returned along the specific route to the destination. Upon arrival at the origin of the ARP request, it caches this information for later use.

ARP responses are normally answered by either the destination node or by a router serving that node. The origin of the request has no idea which is applicable upon getting the response.

4.2.3 Routing

Routes through the IP network are done by first determining the relative position within the network and determining what node has responsibility to reach that destination. The first level of determination is done by reviewing the destination address.

4.2.3.1 Network addresses. An IP address consists of a 4-byte address. In order to provide a more humanly readable address, this is normally shown as a four-component address. Each part of the address is represented by the relative byte of the node's address. Thus, an IP address is represented as an address in the form [a.b.c.d], where each byte's decimal equivalent number is separated by a period.

4.2.3.2 Subnet. IP networks are subdivided into subnetworks.[5] Unlike subarea SNA and APPN, it is possible to look at a network address and determine which subnetwork the node resides within. This can be done by using the *subnetwork mask*. This is a mask that is used by ANDing the destination node's network address to determine if the destination is within the originating node's subnet.

This is done in the following manner:

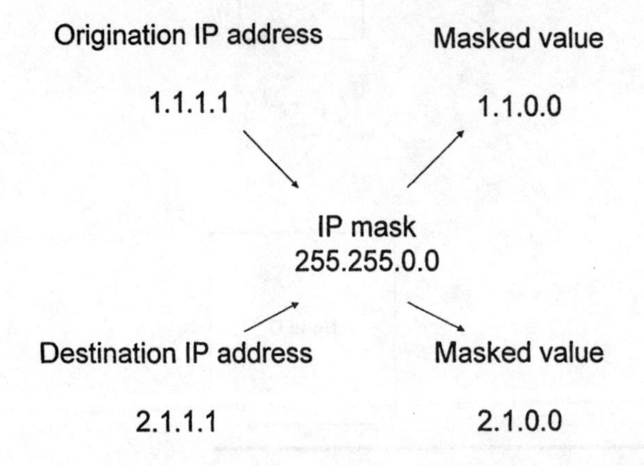

Figure 4.4 Subnet mask.

[5]A subnet is very similar to the concept of a domain within subarea SNA and APPN.

As you can see in this example, originating node 1.1.1.1 immediately knows that 2.1.1.1 is in another subnet by comparing the result of ANDing the local node's address against the result of ANDing the destination node's address. If the result is the same, the two nodes are within the same subnet. If they are different, they reside in different subnets. It is then necessary to determine in which subnet the node resides.

4.2.3.3 Routing table. Each IP node contains a routing table. This table contains the destination addresses and the gateway to that destination.[6] The IP node searches the routing table to locate an exact match of the destination address. If this is not located, then a match for the network of the destination is located.[7] If one is found, then the request is sent to the specified gateway. If a match is not located, then the default entry is used.

4.2.3.4 Routing protocols. In addition to the routing tables, an IP node uses some type of protocol to determine how to reach a destination and to provide information exchange on route and node existence. Two examples of routing protocols are:

- Routing information protocol (RIP)
- Open shortest path first (OSPF)

RIP is the most widespread routing protocol. It has been included in the BSD TCP/IP implementation and is still available as the *routed* daemon. It has also been standardized in RFC 1058. All of the routing protocols are based on the idea of routing metrics. RIP uses a fairly simple idea of routing metrics that is based on a *distance vector*. This is a computed cost for each hop in the network. RIP does have a limit in that the computed metric cannot exceed 15. For this reason, RIP is normally used by assigning a cost of one for each hop. Thus, RIP bases the route based simply by hop count. RIP is also very noisy, in that route information is sent every 30 seconds. If there is a high number of nodes, this background transport of network routing information can clog the network. This is especially true if relatively slow wide area network links are used.

---

[6]Other information may be contained within the routing table, but this is the minimum.

[7]A match is done by using the subnet mask.

OSPF scales much better than RIP. This is because of the more efficient information that is passed between nodes. OSPF also provides support for splitting network traffic across multiple routes, routing based on other metrics than just distance, and the authentication of routing update messages. OSPF gains much of its efficiency by splitting the network into areas. Area 0 is designated to be of special significance because it is the backbone that provides connection between all of the other areas. Each node keeps awareness of its own area. Routes that cross an area boundary is calculated by obtaining only summary information that has the connections to the other areas, not a detailed view of the other area. A node obtains its neighbor's routing table upon connection. From then on, only changes are transmitted.

4.2.4 Transport protocols: UDP and TCP

The customary transport layers used with IP are UDP and TCP. These protocols are used as the interface between an application and the IP network protocol. Each of them has distinct characteristics that make it suited to a specific user requirement.

Both UDP and TCP utilize an addressing that allows an application to direct a message at a specific remote application. This is done by assigning a 16-bit identifier to each message. This identifier, known as a *port,* allows the message to be directed to a specific endpoint.

Ports have been defined to specific applications. These defined ports are called *well-known ports.* These are defined by the Internet Assigned Numbers Authority (IANA) and is published in the Assigned Numbers RFC. Table 4.1 illustrates some UDP well-known ports.

The pool of available ports for UDP and TCP are independent; that is, port 123 for UDP is not the same as port 123 for TCP although addresses are not normally overlapped in this way. This is done for the usability of humans, not because of any UDP or TCP requirements.

4.2.4.1 User datagram protocol (UDP). UDP is a simple protocol that relies simply on passing messages to the IP network layer. Since IP is an unreliable layer and there are no additional error recovery methods imposed by UDP, this protocol is only used when a low-overhead, although unreliable transport, fulfills the customer's requirements.

UDP uses a datagram as the basis for all communication. Each message must be self-contained. That is, it cannot rely upon a reliable transport, and be flexible enough to allow for both lost and duplicated messages. In return, UDP requires very little overhead. As each mes-

TABLE 4.1 Well-known UDP Ports

Service	Port	Description
echo	7	Echo test service
quote	17	Returns "quote of the day"
chargen	19	Character generator
nameserver	53	Domain name server
bootps	67	Server port used to download configuration information
bootpc	68	Client port used to receive configuration information
TFTP	69	Trivial file transfer protocol
SunPRC	111	Sun remote procedure call
SNMP	161	Used to retrieve network management queries
SNMP-trap	162	Used to receive network management problem reports

sage is self-descriptive, there is no error recovery, no sequencing of messages, and no extensive error detection. Because of the low overhead, messages can pass through the UDP layer very quickly. In addition, the IP layer also provides low-overhead layers to provide high throughput.

UDP uses the protocol identifier of 17. When an IP message is sent or received, this protocol field allows the message to be directed at UDP. Further identification is done through the use of the UDP source and destination port numbers and the IP address. The UDP header provides these fields along with a checksum.[8]

4.2.4.2 Transmission control protocol (TCP). Unlike the IP protocols described so far, TCP provides a reliable transport for the use of applications. It also provides a session between the endpoints. This session allows each endpoint to keep track of data flow, sequencing of messages, and error notification to the remote application.

TCP applications pass streams of data to TCP. These streams can exceed the maximum size of the network to be transported at one time. If this is true, TCP will transparently segment the message into pieces that are smaller than the maximum transmission size.

Like UDP, TCP has its own set of well-known ports. Some examples of these are shown in Table 4.2.

TCP uses a request numbering and acknowledgment scheme that is quite simple, but that provides multiple purposes. Rather than numbering each message, as SNA and APPN do, TCP uses the amount of data as the sequence number. Thus, after an initial sequence number

---

[8]The UDP checksum is used to validate the contents of the UDP frame. The checksum is applied to a pseudoheader, UDP header, and the user's data.

TABLE 4.2 Well-known TCP Ports

Service	Port	Description
discard	9	Discards all incoming data
chagen	19	20exchange streams of characters
FTP-data	20	File Transfer Protocol data transfer port
FTP	21	File Transfer dialogue port
Telnet	23	Remote login
SNMP	25	Simple network management protocol
X400	103	Used for X.400 mail services
POP3	110	Used for PC-based mail

is chosen, TCP counts characters on each transmission, adds this to the last acknowledged sequence number, and puts this in the TCP header as the sequence number. This procedure allows not only the ordering of messages, but is a simple method of ensuring that all of a message has arrived and to detect the duplication of messages.

4.3 TCP/IP Formats

Each of these layers uses its own headers. Because TCP is a layered architecture, if you took a snapshot of a frame actually being transported, you would see that the higher layer headers are embedded within the lower layer frames. Thus, a TCP frame is embedded within an IP frame, which is inside of the media access control (MAC) frame. Figure 4.5 shows how each of the layers pass the frame down, where an additional header is added. Upon being received, each layer processes its own header and passes the frame, minus that layer's header, to the next higher layer.

4.3.1 IP header

Figure 4.6 shows a view of the format of an IP header. The header contains several fields that are used to either specify addressing of the frame (source and destination addresses), the transport protocol of the frame (UDP is 17 and TCP is 6), and some control information.

The IP header contains a checksum field. This checksum is recomputed for each hop through the network. Fields such as time-to-live, fragmentation, and options are included in the checksum calculation.

4.3.2 UDP header

Figure 4.7 shows the format of the UDP header. This header, as this transport layer exhibits, is simple. It contains only the source and

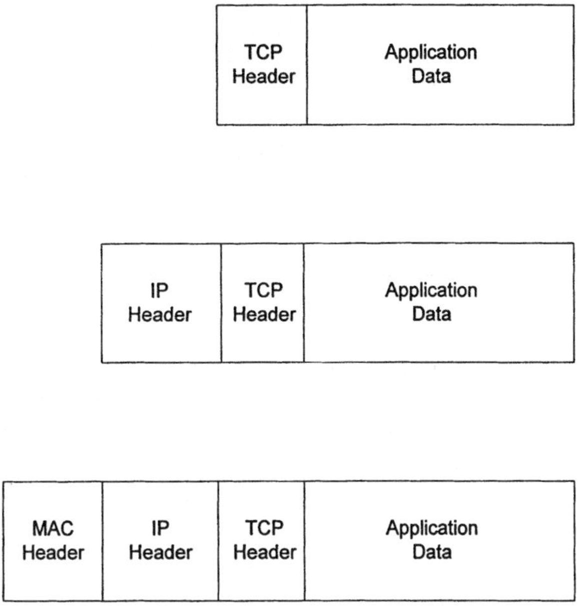

TCP Header	Application Data

IP Header	TCP Header	Application Data

MAC Header	IP Header	TCP Header	Application Data

Figure 4.5 Addition of headers to message.

<table>
<tr><td>Byte</td><td colspan="2">0</td><td>1</td><td colspan="2">2</td><td>3</td></tr>
<tr><td></td><td>Version</td><td>Header length</td><td>Type of service</td><td colspan="3">Total length of datagram</td></tr>
<tr><td></td><td colspan="3">Identification</td><td>Flags</td><td colspan="2">Fragment offset</td></tr>
<tr><td></td><td colspan="2">Time to live</td><td>Protocol</td><td colspan="3">Header checksum</td></tr>
<tr><td></td><td colspan="6">Source address</td></tr>
<tr><td></td><td colspan="6">Destination address</td></tr>
<tr><td></td><td colspan="6">Options
Strict source route
Loose source route
Record route
Timestamp
Security
Padding</td></tr>
<tr><td></td><td colspan="6">User data</td></tr>
</table>

Figure 4.6 IP header.

<table>
<tr><td>Byte</td><td>0</td><td>1</td><td>2</td><td>3</td></tr>
<tr><td></td><td colspan="2">Source port</td><td colspan="2">Destination port</td></tr>
<tr><td></td><td colspan="2">Length</td><td colspan="2">Checksum</td></tr>
</table>

Figure 4.7 UDP header.

Byte	0	1	2	3
	Source port		Destination port	
	Sequence number			
	Acknowledgement number			
	Data offset	Flags	Window	
	Checksum		Urgent pointer	
	Options and padding			
	User data			

Figure 4.8 TCP header.

destination ports, a length field, and the checksum. In addition to this header, there is a pseudo-IP header. This pseudoheader is a scaled-down version that provides only some of the IP header fields. Since no error recovery, sequencing, or other more complex functions is provided by UDP, the header does not contain any place to specify these.

4.3.3 TCP header

The TCP header is shown in Figure 4.8. Because of the more extensive services provided by TCP, its header is more extensive than that of UDP, including such fields as:

- Sequence and acknowledgment numbers
- Window
- Flag bytes for session establishment and maintenance

TCP must manage the session between the endpoints and thus requires additional fields. The use of the flag byte allows session partners to pass status information to their remote partner. This is used to maintain the session and to ensure that the session partners are synchronized.

4.4 Communication Software

TCP/IP is one of the most widely supported network architectures today. The protocol is supported by computers ranging from hardware components to the largest supercomputer. Because the TCP/IP stack[9] is small and simple, it has been made available for components

[9] A protocol *stack* is the software required to implement the protocol.

through ROMable software. This allows individual components to communicate their status information to a specific destination.

UNIX operating systems have a native TCP/IP implementation that is included with the operating system. Because of the widespread use of UNIX, TCP/IP is supported by a wide range of systems throughout the world.

There are also PC-based implementations of this protocol. These usually operate as a series of device drivers, which usually control the operation of a network interface, and some kernels and applications. These allow full TCP/IP use and, as such, allow communication with any platform that supports TCP/IP.

4.5 IP Applications

As has been presented, IP has some well-known ports. Each of these ports directly corresponds to an application. Although every well-known application does not have to be present, it is not advisable to use a well-known port for a user-written application.[10] Among the well-known applications are:

- File transfer protocol (FTP)
- Telnet
- TN3270
- Simple mail transfer protocol (SMTP)

4.5.1 File transfer protocol (FTP)

File transfer protocol (FTP) is one of the most widely used applications in the TCP/IP suite. It is used to allow files to be transported across an IP network.

To allow the maximum amount of flexibility, FTP can provide file services for either a character or binary mode. The character-mode is used to provide some allowance for conversion to and from different character formats, such as between UNIX and DOS systems. FTP can add or remove the carriage-return (X′0D′) code when moving between a UNIX and DOS system.[11]

---

[10]Actually, nothing adverse occurs, except that someone making a request to a standard application may not obtain the service that was requested.

[11]DOS uses a carriage-return (CR), line-feed (LF) (X′0a0d′) combination for each line of characters. UNIX files contain only a LF after each line.

Binary mode is used when noncharacter data must be transferred. In this case, data is transported through the network in a way that results in no bit modifications of the data. This provides the ability to pass data without regard to the format of the data. For example, a DOS file can be placed on a UNIX fileserver. This data can then be transferred to another DOS system without the data conversion that results from a character transfer.

4.5.2 Telnet

The *telnet* application allows a client system to log onto another system as a terminal. Thus, a client can access a remote system but act as an attached terminal to the remote system. The application that you are accessing is operating on the remote system; the telnet client merely allows the remote access. In most cases, the telnet client acts like a VT100, as this was the most common terminal at the time this program was written.

Telnet connections use some control characters, just like a real terminal. These control characters are the basic American Standard Code for Information Interchange (ASCII) control characters, such as tab, form feed, bell, and several others. There is also a telnet escape sequence that allows you to "escape" to the telnet control mode.[12]

Telnet options can be negotiated between the client and server. Some of the options include:

- *Binary transmit.* When *binary transmit* is used, data is transferred as 8-bit data. If the escape sequence is to be sent as data, it must be sent twice; the first sending puts the session into escape mode, while the latter is taken as data.
- *Terminal type.* Many different types of terminals can be emulated. The set that is available is specific to the TCP/IP implementation that you are using.
- *Echo mode.* Just as terminals can specify that the server echo characters back, so they can be displayed on the screen, telnet also allows this same option.
- *Virtual screen size.* The virtual screen that is used can also be negotiated. This allows extended screen sizes to be fully used.

---

[12]The default escape sequence is ^] (control]).

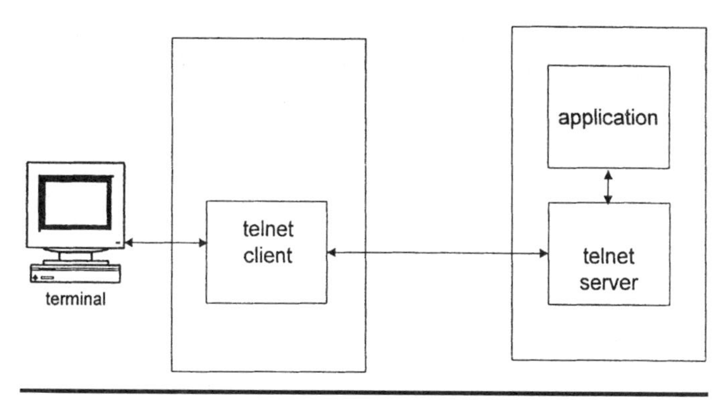

Figure 4.9 Telnet dataflow.

4.5.3 tn3270

Access for IBM-type terminals may be implemented within the telnet command, but is normally done through a separate program called *tn3270.* This program uses the binary mode and End-of-Record codes to allow 3270 datastreams to be sent to a terminal type that is not normally utilized, such as VT100. In this case, the telnet server acts as the end device to the 3270 application. Codes are modified to allow the data to be used on the nonnative terminal device.

Figure 4.9 shows how the telnet server and client reside in the network. The telnet server operates as a terminal to the application. This data is transported across the network to the telnet client, which reformats the data for the specific terminal being emulated.

4.5.4 Simple mail transfer protocol (SMTP)

SMTP is the standard method for passing electronic mail between systems. SMTP is the background process that allows the mail messages to be transferred; it is separate from the agent that users use to enter, read, and manage mail. A mail message consists of two components—the message and the mail address. The separation of these components allows SMTP to transfer messages without having to scan a message to extract an address.[13]

Similar to IP addresses and a domain name server, the address is used to determine if the destination is local or remote and how to

[13]There is another standard for mail. This is the OSI standard X.400. There are many gateways between SMTP and X.400, as well as gateways to other mail message types.

route the message to the destination. In the case of SMTP, addresses consist of a user and domain name. For example, *user@this.domain* can be separated at the "@" to user, the user component, and *this.domain,* the domain name.

In many cases, the domain name component consists of a hierarchy of areas. For example, *mary@phx.mycompany.com* creates a name tree of:

TABLE 4.3 Breakdown of SMTP Name

com	Commercial
mycompany	Name of commercial entity
phx	In the city of Phoenix
mary	Name of user at this location

This domain name is used to determine how to route the message.

4.6 Summary

The IP architecture was derived from some early work conducted by the DOD. The objective of this work was to create a network architecture that was highly flexible and could recover from almost any message failure. It also had to be highly dynamic in both its network routing and in operation; single-point of failure was not conducive to the objective of survivability.

The architecture was created by academic centers across the country. These same universities and research centers were concurrently working with a new operating system called UNIX. In an effort to create usable implementations of the new architecture, the protocols were created under the UNIX operating system. As UNIX was distributed across the country, the TCP/IP implementation went with it. This lead to the quick acceptance of TCP/ IP. This acceptance was furthered by the fact that the network that these same people were using utilized TCP/IP, as well as the fact that the implementation was free and full.

The TCP/IP protocol stack was based on the military objective of survivability. This survivability objective lead to the design of a network layer that did not provide recovery and that network errors could only be recognized by the endpoints of the network traffic.

Routing through IP networks was dynamic. Each frame was independently routed through the network. Although the network learned about the available paths and algorithms that were developed to optimize routes, frames still often were lost in the network. Recovery of these frames is the responsibility of the higher layers, such as UDP and TCP.

IP networks developed not only the network architecture, but also some "standard" applications, such as file transfer and remote logon. These applications were standardized through a consensus. As these applications were developed, they also became a component of the codebase for TCP/IP.

Chapter

5

Comparison of Network Architectures

The four network architectures (i.e., subarea SNA, APPN, HPR,[1] and TCP/IP) dealt with in this book represent a large portion of the networks that are currently operating.[2] These architectures also transport the vast majority of network traffic throughout the world.

As was previously explained, each of the architectures represents a different view of the responsibility of the network components to the health of the network. Subarea SNA, APPN, and HPR are similar in their view of network responsibility, but there are still definite differences among them. Internet Protocol embodies a very different view. As was discussed in Chap. 4, IP is based on the belief that reliability is dependent on the actions of the session endpoints.

Prior to investigating how these architectures can be integrated, a comparison of the various network architecture to each other is in order. I will set forth a short but concise comparison of each network type, the assumptions that it makes, and how common events are acted upon.

5.1 Philosophy of the Network Components

Just as we react to events around us based on our assumptions about the world, networks also exhibit similar behavior. Networks rely upon

[1]Because HPR is based upon APPN, it has all of the attributes of APPN. As a result, I will address only the HPR characteristics that set it apart from APPN.

[2]HPR is only now becoming available from a select set of vendors. As a result, there is almost no market penetration by this architecture.

their assumptions to determine how to react to events. These events include, but are not limited to:

- Loss of a communication path
- Loss of data frame
- Retransmission of an incorrectly received frame

Subarea SNA and APPN view the network as unreliable. The paradigm for these network architectures is that one of the main purposes of the network is to create a reliable foundation upon which session partners can exchange data. To accomplish this, each of the links within the network provides functions that enable the foundation to appear completely reliable. These functions include the detection of line errors and the retransmission of frames. These retransmissions are repeated until either a frame is successfully transmitted or a specified limit is reached, at which time an error is reported to a higher level.

The link-level error recovery tower of HPR views communication links as largely reliable. As a result, the retransmission of frames is pushed to the endpoints of a connection. Communication errors are recognized and corrected only at these endpoints. Intermediary nodes use the ANR component of the HPR architecture to pass frames. As a result of the reduced communications overhead at each intermediary node, higher speed communication paths can be effectively utilized.

Although error detection is moved to the periphery of a connection, HPR still creates a reliable transport for higher-level SNA services by allowing endpoint error detection and correction. By providing a reliable communication foundation, such components as session services can know that communication is successful. Session services rely on the reliable transport to pass data on sessions between LUs.

The foundation of TCP/IP, the IP layer, is based on the paradigm that the network is not responsible for error detection and recovery. This is the realm of endpoints for data transmission. By isolating itself from these responsibilities, IP provides a fluid network that sends datagrams between points within the network. Since the network is not responsible for error detection, the complexity of each node within the network is reduced. Each node within the network only has to be able to make some simple routing determinations and transmit data. Buffers are reduced because there is no requirement to save data for retransmission. This greatly reduces the cost of creating nodes within the network.

5.2 Layered View

This section compares the four network architectures that are the subject of this book: subarea SNA, APPN, HPR, and TCP/IP. The purpose of this analysis is to briefly review how each of these architectures views each of the OSI layers, how they are similar, and how they differ. Through this analysis, we will be better prepared for the rest of the book, which describes methods of integration.[3]

5.2.1 Physical layer

The network architectures we have reviewed do not have any physical layer differences. In fact, all of them support the same physical interfaces.

Subarea SNA, APPN, and HPR are often viewed as supporting only serial communications. This is because they are normally operating through a communications controller, such as a 3745. Although this type of controller supports LAN interfaces, the historical view of these connections as being serial-based still pervades the view of users.

However, in the last few years more users of SNA-based networks have started using high-speed LAN interfaces. This has led to a new industry in conversion from remote serial communications to LAN interfaces.[4] There is also some movement toward channel interfaces, to obtain the higher throughput enabled by this type of interface.

There has also been increased interest in alternative communication methods. These include several high-speed alternatives such as ATM and SONET. Alternative lower-speed serial methods, such as frame relay, are also of high interest to users throughout the world. Most of these expect that this type of endpoint error recovery will be utilized. This results from the incompatibility of higher-speed communication interfaces and redundant error recovery.

TCP/IP is often viewed as being synonymous with ethernet LANs. Just as the SNA architectures contain no inherent limitation, there is no architectural limitation. TCP/IP also supports several serial implementations, such as Point-to-Point Protocol (PPP) and Serial Line

---

[3]It should be understood that at the physical and link layers, all of the architectures have no inherent limitation in their support of almost any implementation.

[4]These LAN interfaces are normally token ring, as this is the most common environment to an SNA host. Often this connection is through a token ring connection to a communication or cluster controller.

Interface Protocol (SLIP), but LAN connections are seen as the normal implementation.

5.2.2 Link layer

The link layer specifies the protocol that is used to ensure that messages are sent and received correctly across a single physical interface. The purpose of this layer is to provide an error-free foundation upon which higher layers can operate.

As was previously stated, the link layer is largely open in each network architecture, though some link layers are seen as synonymous with certain architectures. Specifically, the Synchronous Data Link Control (SDLC) link protocol is seen as being synonymous with subarea SNA and wide area connections for APPN and HPR. Although these network architectures actually support almost any link protocol, since SDLC was designed by IBM for SNA, this view is not without some basis.

Likewise, carrier sense multiple access with collision detection (CSMD/CD), which is the link layer for ethernet, is often seen as being synonymous with TCP/IP. Since many TCP/IP installations use ethernet, there is also reason behind this view. But, as was true of SDLC and SNA, this is not actually a limiting factor. You can run TCP/IP over a SDLC link layer or operate SNA over an ethernet connection.

5.2.3 Network layer

The network layer provides for end-to-end routing of data. Since SNA, APPN, HPR, and TCP/IP have different methods of routing traffic, this is the first layer that provides for some contrast between the network architectures.

5.2.3.1 Network layer—subarea SNA. Subarea SNA provides a network layer that is very robust and fast. Because the routing tables for subarea SNA are predefined, little overhead is associated with determining routes. Routes between nodes are fixed within the defined tables.

The overhead of creating these tables is not eliminated, rather it is moved to the system programmer who needs to develop the table. Although the network does not experience overhead by having to determine routes, a high price is paid by the person that must create these tables.[5]

5.2.3.2 Network layer—APPN. A major reason users of subarea SNA upgrade their networks to APPN is because the routing tables are created dynamically by the network, instead of the system programmer. As was just discussed, the time required to produce the network routing tables for subarea SNA is high. In addition, because routes need to be defined to all destination domains, the tables must either contain entries for domains that do not exist yet or must be updated to provide routes to new domains.

APPN removes all of these predefined routing tables. Instead, the APPN node learns dynamically about the network topology and calculates routes according to that topology. Thus, routing tables can be built by the network rather than the system programmer. Such a process provides for the development of an accurate topology, because it reflects the actual network. As nodes are added to the network, they are reflected in the topology tables of each APPN node automatically.

Because APPN has a foundation of a session, the components build a base of a reliable transport. The use of the APPN path control component provides for this base. This contrasts with the datagrams associated with IP, as seen in the next section.

5.2.3.3 Network layer—HPR. HPR extends the network layer of APPN with ANR. When an RTP connection is being used, ANR reduces the processing required for intermediate routing nodes.[6] This is done by creating a source route that is similar to that used by token ring nodes. Because the route is established on activation of the route, the processing required by intermediary nodes is reduced[7] while actually routing traffic.

The RTP connection across an HPR network forms a virtual pipe across the HPR network. During the setup of this connection, a route is calculated and ANR labels created by each node along the path. In addition to route determination, a determination of the maximum

---

[5]For more information on this topic, see Sec. 5.3.1.

[6]If no RTP connection is being used, HPR nodes provide the same interface as any other APPN node. There are no advantages or differences between HPR and APPN nodes in this configuration.

[7]This processing reduction is in addition to the elimination of intermediary error detection and recovery.

segment size is made so that no intermediate node needs to provide segmentation.

A series of ANR labels is determined as the source route is built by the nodes along the route. These ANR labels can be used to assist in speeding up the lookup of the next hop along the path. For example, these labels may be control block addresses that allow almost immediate routing of packets across the HPR network.

Another key to ANR is that intermediate nodes have no session awareness. Session traffic flows through intermediary nodes without these nodes having to participate in any session knowledge. This flow-through manner speeds the process of routing while reducing the overhead on these same nodes. This win-win situation results in an effective and efficient network layer.

5.2.3.4 Network layer-TCP/IP. TCP relies on the dynamic routing feature of IP to provide a network layer. The attractive property of this network layer is that IP uses a hop-by-hop routing scheme that provides an extremely flexible network layer. Although messages may be duplicated or arrive out of order, this situation is rectified by the next layer, the transport layer.

The line overhead that can result from the duplicated messages can be a problem, especially when a low-speed WAN link is used. This can be an issue to a user, but is not one for the network layer. If a series of messages must be retransmitted by an endpoint through the network, the additional data traffic has an adverse effect on all traffic throughout the network.

5.2.4 Transport layer

The transport layer provides the assurance that data successfully moves from origin to destination. The responsibilities of this layer include the regulation of message transmission, the resequencing of messages into their correct order, and ensuring that the data has been successfully received.

5.2.4.1 Transport layer—subarea SNA. Subarea SNA provides a robust transport layer. By using the features of the DLC, link, and network layers, subarea SNA provides a reliable foundation for the transport layer to implement a reliable transport.

All information frames include a sequence number that allows the endpoints to determine if messages arrive in their proper sequence. If messages do not arrive in their proper sequence, subarea SNA nodes signal for a retransmission of missing or garbled frames.

Subarea SNA also allows messages to be packetized into smaller pieces called *segments*. This enables large messages to be broken down into more manageable sized packets that move through the network easier. When these packets arrive at their destination, they are put back together in their original format so that the complete message can be used.

5.2.4.2 Transport layer—APPN. APPN is built directly upon the transport layer of subarea SNA. All of the functions that exist within the subarea SNA transport layer also exists within APPN. This allows for a smooth migration and integration path for these two architectures.

In addition to the facilities of subarea SNA, APPN also supports the segmentation of some control messages, specifically, the BIND. This is required because the BIND in APPN can grow substantially due to the large number of control vectors that are included in APPN communication. By being able to support segmented BIND messages, the network is better able to support the full range of functions that APPN can facilitate.

5.2.4.3 Transport layer—HPR. If no RTP connection is being used, HPR nodes provide the same processing support and interfaces as a base-level APPN node. Differences appear only once an RTP connection exists.

In this case, HPR nodes rely on the RTP connection endpoints to provide error recovery processing. Intermediary HPR nodes utilize a reduced link-level error recovery stack that does not provide this service to traffic flowing through them. The elimination of error recovery procedures from these nodes greatly reduces the processing overhead on these nodes. At the same time, it allows better support for high-speed connections. The elimination of these error recovery procedures ensures that packets flow through these nodes without impediment. There are no retransmission of frames or link-level acknowledgments that could slow the flow of traffic along the route.

The error recovery procedures are moved to the endpoints of the RTP connection. These endpoints provide the key error recovery procedures for traffic flowing through an RTP connection. If an error is detected, an endpoint signals a request for retransmission to its partner endpoint.[8] The packet is retransmitted across the entire RTP connection.

---

[8]As was stated earlier, this connection does not constitute a session.

Although error recovery is moved to the connection endpoints, this architecture assumes that the links being utilized have a high reliability profile. It is because of this profile that retransmissions across the entire RTP connection do not adversely affect data throughput.

The RTP connection endpoints also support selective retransmission. This allows for the retransmission of only a portion of the transmission window. For example, if a transmission window is modulo 128 and all but the last packet is received successfully, the RTP endpoints can signal that only the last packet be retransmitted. This further reduces the impact of retransmissions across the length of the RTP connection.

An RTP endpoint can also recalculate and use the new route without requiring that the endpoint applications/LUs be notified or modified. This route recovery is instituted by one of the HPR endpoints recognizing the failure of a segment along the RTP connection route. Once detected, a new route is calculated that uses only HPR nodes to the same destination. Assuming that a new route can be calculated, data is moved to the new route without requiring any endpoint notification.

5.2.4.4 Transport layer—TCP/IP. Internet protocol does not provide a transport layer. There is no detection of any missed frames, overrun frames, or an incorrect sequencing of messages. This architecture does not provide any of these features.

However, TCP does contain these features, but does not have the functionality to implement most of them. This is a result of the datagram basis of IP. Because there is no reliable foundation, TCP must provide all of the transport layer itself.

5.2.5 Session layer

The session layer provides for the formation of a simulated direct conversation between entities within the network. This conversation allows for these points in the network to pass information "directly" between them.[9] Message sequencing and delivery confirmation are components of the session layer.

5.2.5.1 Session layer—subarea SNA. Subarea SNA was designed to enable data transmission from dumb remote devices to a centralized data center. One of the assumptions of the designers was that com-

[9]"Direct" means that a direct connection is simulated by the network. Though they are usually not actually adjacent, the network provides the illusion that they are by providing a strong foundation.

munications was unreliable, i.e., that lines were unable to transmit data correctly.[10] Line errors were common and messages were known to get lost within the network.

To overcome the view that communications could not be relied upon, the designers of subarea SNA built features that increased both the reliability of the network and error detection. To this end, a new data link control, SDLC, was developed. This DLC had a more robust ability to detect line errors through the use of a more reliable detection scheme than existing DLCs. SDLC also provided for message acknowledgment to the DLC partner and allowed for the signaling of a retransmission.

Subarea SNA is based on the session, i.e., every message in subarea SNA is sent across a session. Sometimes these sessions are between entities that are nonadjacent. In some cases, the messages are passed to sessions within a single physical component.

Subarea SNA defines several different session types, which include:

- SSCP-SSCP
- SSCP-PU
- SSCP-LU
- LU-LU

Each of these sessions provides different functions and handles different types of data. These distinctions provide for standardized interfaces between components in the network.

The SSCP-SSCP session communicates between distinct VTAM systems. The type of information that is passed across this session are session requests that cross domain[11] boundaries. These sessions also coordinate some control functions between these same VTAM systems.

The SSCP-PU session provides control functions within a single domain. The control functions include the management of subordinate entities, LUs, and subordinate sessions, SSCP-LU and LU-LU. Because subarea SNA is a hierarchical architecture,[12] each component in the hierarchy must be active before any subordinate entities or ses-

[10]Communication lines are called "dirty."

[11]For information on this term, see Chap. 2 or the Glossary.

[12]Subarea SNA was based on the idea that communication was between terminals and mainframes. The host is seen as the destination of all traffic in the network. From this point, the NCPs exist with their lines. Subordinate to the lines are the PUs or cluster controllers, and LUs are further subordinated.

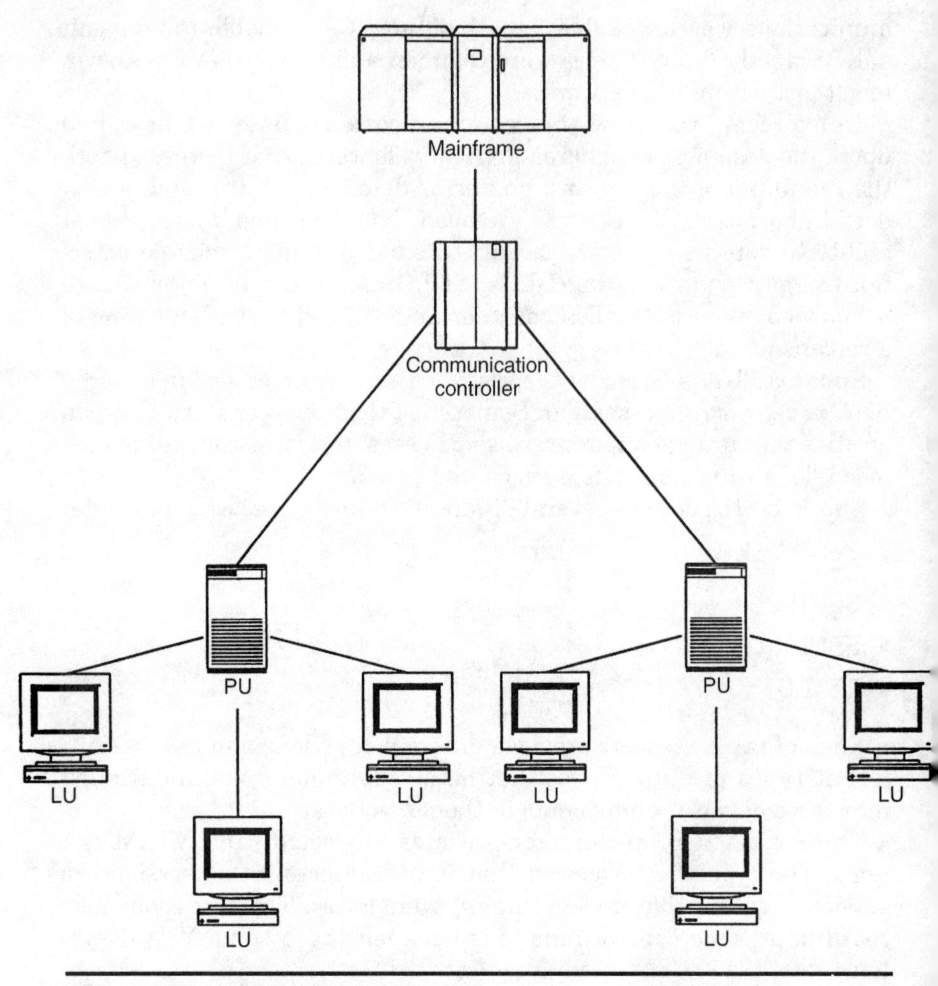

Figure 5.1 Hierarchical network.

sions are activated. Thus, the PU must be active before activating any of the LUs associated with that PU. Figure 5.1 shows this hierarchy.

The SSCP-LU session provides the communication upon which the most important session, LU-LU, is based. The SSCP-LU session enables the SSCP to gain awareness of the LU and for the LU to obtain the ability to request a session. Thus, the SSCP-LU session is beneficial to both the SSCP and the LU in that it helps them obtain the services of the partner.

The LU-LU session is the only session that provides actual user processing. This is the session for which all of the other session exist. Users are only aware of this session, unless a session component fails, at which time they become aware only that they cannot get their work done.

A session is established by the transmission and response to a BIND request. The BIND request specifies the LU address of the primary and secondary LU partners. It also contains session parameters, such as session-level windows, message size limits, and several session-level parameters. Within subarea SNA, the primary LU partner sends a nonnegotiable BIND to the secondary LU partner. The secondary LU has a choice of either accepting the session parameters or it may reject the BIND.[13]

The LU partners number all user data requests that pass separately through the network. Each LU-LU session maintains its own numbering scheme. It is the responsibility of each LU partner to maintain its own session state and sequence numbers.

Sequence numbers allow an LU to both recognize the loss of a message and allow, optionally, an LU to resequence messages into their proper order. These sequence numbers also allow the LUs to stay synchronized during message transmission.

LUs can also provide message delivery confirmation. The confirmation can be done by sending either a definite or exception request. A definite request provides for the explicit confirmation of a delivered message to the partner LU. Upon receiving a request that contains a definite response request, the receiving LU responds with a positive or negative acknowledgment of message arrival.

An exception response requires either an explicit negative response or an implied positive response, i.e., a negative response is *not* sent. The sender of an exception response receives an acknowledgment message from the receiver upon receiving the exception response when the transmission window is closed. Thus, if the window is set to 3 and the first request was numbered 1, the sender of the exception response request receives an acknowledgment message from the receiver upon receiving the first three messages. In this case, the sender would receive three responses for definite response request mode and only one response for exception response mode.

For the PU, a transmission window is established that provides a limit on how many messages are transmitted before an acknowledgment is needed. The message can be used for multiple associated LUs. This windowing allows for better throughput from the network.

Session-based windows can operate by using *pacing*. The purpose of pacing is to prevent a high-speed LU from overrunning a slower LU.

---

[13]Negotiable BIND is supported for LU 6.2 sessions.

The slower LU paces the high-speed LU by withholding pacing acknowledgments. This causes the sender to stop transmitting until a pacing response is returned.

Subarea SNA also allows for messages to be packetized at the session level. This is called *chaining* and is very similar to the SNA segmentation method of the transport layer. The session partners determine the largest packet that can be transmitted. The session partners then chain so that the entire message can be transmitted. This is illustrated in Fig. 5.2.

Subarea SNA allows session partners to specify how communication is to operate. This includes rules on which partner can transmit data without requesting permission, how conversations operate, and recovery responsibilities.

5.2.5.2 Session layer—APPN. APPN was designed to allow communication points in the network to establish a session without any host-based assistance (SSCP). Unlike subarea SNA, APPN nodes do not require the intervention of an SSCP to provide session services. Instead, LUs are able to create, transmit, and respond to session services. In addition, an SSCP is not required to establish a route between LUs.

Like subarea SNA, APPN is also a session-based network architecture. All communication within this network architecture is session oriented.[14]

In contrast to subarea SNA, APPN only supports LU-LU and CP-CP sessions. Since there is no SSCP in APPN, there are also no SSCP-

[14]At present, there is no method for datagram transmissions.

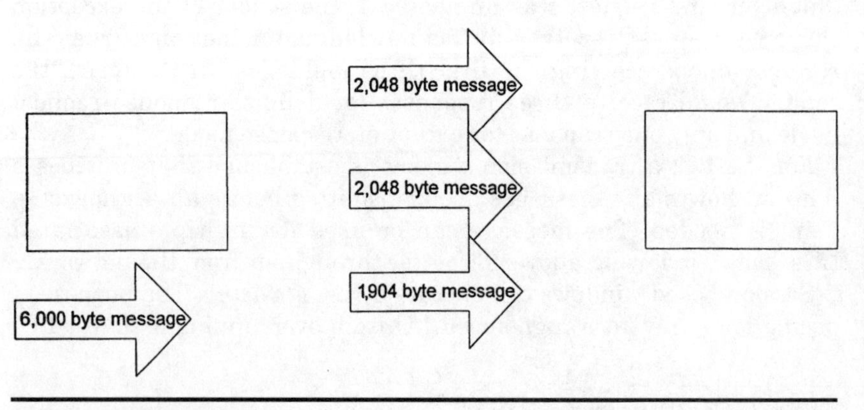

Figure 5.2 Chaining of message.

related sessions. Thus, the only subarea SNA session type that APPN also supports is LU-LU.

APPN adds a new session type called the CP-CP session. This session is very similar to the SSCP-SSCP session of subarea SNA, in that it is the basis of control for the network.[15] The CP of APPN is a specialized LU that provides management services for the node and establishes a session with the CP of another APPN node.

Although sessions are established directly from any APPN node, a NN is necessary to create a route between the APPN nodes.[16] An APPN EN does not possess the required component to calculate a route.[17]

A session is started using the same BIND command as in subarea SNA. What differentiates APPN from subarea SNA is that before the session is established, the destination LU is located dynamically and the network node or network node server calculates the route for the session. By locating the destination LU dynamically, predefinition is not necessary. Instead, the originating LU sends a request through the APPN network to locate the destination. This request can use either the directed request method, in order to confirm a cached location, or the broadcast request method. The broadcast request method can result in several responses.

This CP-CP session is an LU 6.2 session.[18] This means that the basis of the session is a fully available stack that does not require the development of many new functions.

All of these responses are passed on to the NNS of the originator. This NNS processes the responses to determine the correct path to the destination. Once the calculation is completed, the correct path is returned to the originating node, which sends a BIND to the destination through the calculated session path.

This BIND command contains additional control vectors that further define the session request. If the destination approves of the session request and the suggested parameters, it returns a positive response. This completes the session setup for APPN nodes.

The session route is always calculated at the origin, or source, of a session. In addition, once a session is initiated, the route is fixed. If a more appropriate session path becomes available, the session does not

---

[15]Note that the "CP" in SSCP means control point. The CP-CP session can be thought of as a special SSCP-SSCP session.

[16]An EN can establish a session if a session route is already cached or is local, but an NN has to be available to create the original route.

[17]A session can be established if a route is already known, such as a local resource or a resource on an adjacent node.

[18]All independent LU sessions are LU 6.2.

automatically utilize the new path. This type of session switch is only initiated as a result of an outside event, such as the intervention of a human operator, or if a component of the original path becomes marked as unavailable. Thus, the failure of a session component can actually result in the calculation of a better session route.

5.2.5.3 Session layer—HPR. HPR provides the same session layer as base-APPN nodes. No sessions are supported by HPR nodes that are not supported by base-APPN nodes.

At the same time, there are connection-oriented services extending beyond those supported by APPN. These include RTP endpoint connections and the route setup protocol.

A new extension to the APPN architecture, HPR, allows for dynamic session rerouting. HPR uses a different technique to define a session path. Instead of calculating a session path for each session, HPR uses a new foundation session that exists between a pair of APPN nodes. This session, known as an RTP session, becomes the actual foundation for LU–LU session between nodes. This enables faster session setup, if the new session has requested an HPR-capable route, because session route does not need to be calculated. Instead, the session uses an existing foundation session. Figure 5.3 shows how this operates.

Another outcome of this architecture is that the HPR nodes at each end becomes responsible for the maintenance of these underlying sessions. If a failure is detected, the HPR nodes automatically reestablish the session along a new route. As the LU-LU session operates along these foundation sessions, if the foundation sessions move, the application sessions move along with the RTP foundation sessions. Thus, nodes become aware of communications failures and dynami-

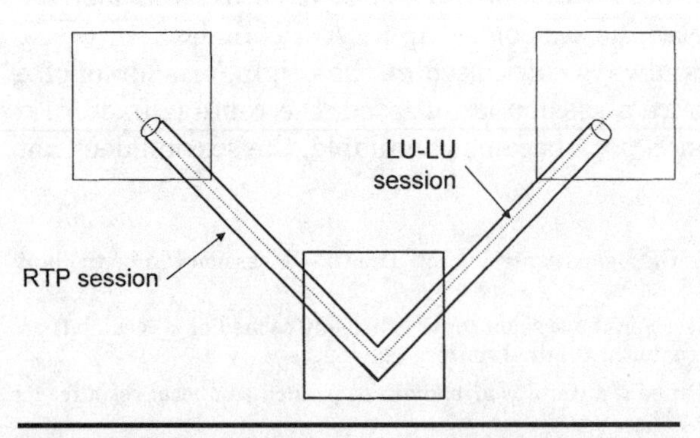

Figure 5.3 High performance routing configuration.

cally adjust the application sessions without informing the session partners.

RTP connection endpoints do not actually establish a session, but they do communicate. This communication is done through the use of datagrams[19] between the endpoints. The use of datagrams reduces the overhead associated with this communication in contrast to the use of a session. At the same time, because these nodes already provide error detection and recovery, there is little additional overhead in the communication.

The RTP NCEs pass sequence numbers, acknowledgment indicators, and other information so that each has a good understanding of the status of communication between them. This information is used for error detection and recovery.

Without the use of a session, no session control blocks are created for the endpoints. Instead, existing status control blocks are used to provide efficient and effective communication.

The route setup protocol is a method of building the route for an RTP connection. The protocol provides for message passing between adjacent nodes in an effort to build the ANR labels that are used along the RTP connection for routing. This protocol is even further from a session than the RTP endpoints, but it does require that adjacent nodes communicate and that information be passed along to the next node, until the destination is reached.

5.2.5.4 Session layer—TCP/IP. Internet protocol uses datagrams to transport data, but does not provide any guarantee or assurances that data handled by IP will arrive at the destination. This function is provided by the TCP layer.

TCP establishes a session with a partner. The session is set up by at least one of the partners sending a SYN command. This command includes some session parameters, such as window size, maximum segment size, and maximum message size. The destination, upon receiving the SYN command, responds with a SYN ACK command. The originator, after receiving the SYN ACK command, returns an ACK command to complete the session setup sequence.

Initial sequence numbers are exchanged during the session setup sequence. After accepting these initial sequence numbers, each

[19]Messages are actually passed on a session between the LUs. These are "datagrams" because there is no link error detection, but only end–end recovery. This use of datagrams for the transmission of data between nodes is a very unusual event for SNA/APPN networks. Although they are datagrams, the data still contains information that allows the two ends to communicate as to their current status.

request updates the expected sequence number by adding the byte count in the message. The window size is also updated to manage flow control.

However, it is the responsibility of TCP, and not the underlying protocol, as in subarea SNA and APPN, to provide the session services and data delivery checks. This results in a higher level needing to be driven to perform these tasks. Because the overhead of IP is very low,[20] data traverses the network easily, but normal session requirements, such as data delivery checks, must be done after passing to a higher layer than in the SNA architectures.

There is some gain and loss by using this method. The overhead of the network is lessened *for each packet,* because there is no error checking. If a message needs to be retransmitted, it will be detected only by the session endpoints. The request for retransmission could be the result of a checksum failure or a missing message. Messages may also be duplicated. This results in the TCP layer having to provide more services than are required in the SNA architectures. On the other hand, each node must make a routing decision for each packet.

5.2.6 Presentation layer

Subarea SNA, APPN, and HPR provide some presentation services for applications. These services include support for several different datastream types, including 3270, 5250, and generalized datastream (GDS).

TCP/IP does not define presentation layer protocols. It is left to the endpoints to provide data in a useful format.

5.3 Definition

Every network architecture requires some level of network definition. At a minimum, a node must at least possess a definition of itself. This definition may include as little as a name and some rudimentary information, such as defining the node's physical connectivity.

The four network architectures have different definition requirements. Some of this distinction is caused by the respective age of each of the architectures; newer architectures tend to require less definition.

[20]This overlooks the route calculation at each node along the path.

5.3.1 Definition—subarea SNA

Of the four network architectures, subarea SNA requires the most extensive amount of network definition. All local and remote devices and lines must be predefined before this architecture is able to develop some dynamic knowledge of the network. In addition, the dynamic portion is limited to cross-domain resources that do not require outbound sessions to be established to them (rather than incoming sessions). Thus, a remote resource that wants to come into the local domain can operate without a definition, but local resources may not access another remote resource without predefinition. Figure 5.4 shows the differential in definition requirements for each direction.

5.3.2 Definition—APPN

APPN built upon this subarea SNA base, but the APPN designers determined that the definition requirements of subarea SNA were too onerous. As a result, APPN nodes need to define only themselves and the physical interfaces to nodes. There is no need to define where a destination is located within the network. APPN nodes dynamically find destinations through the use of a broadcast command that

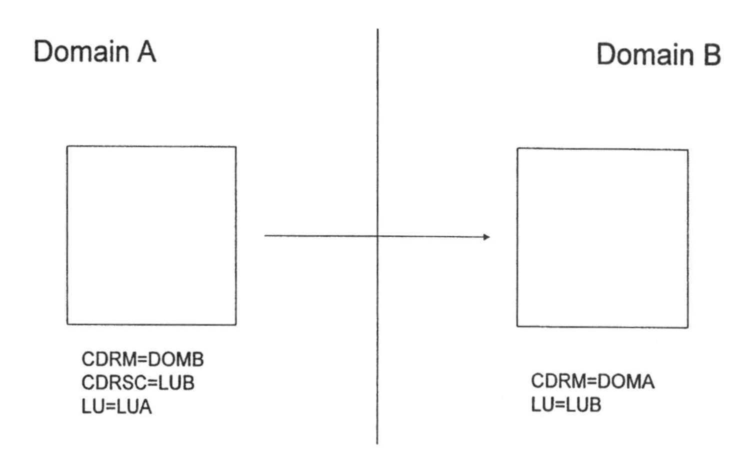

A session from LUA in Domain A to LUB in Domain B can proceed without a definition for LUA in Domain B.

LUB cannot establish a session with LUA if LUA is not defined in Domain B.

Figure 5.4 Subarea SNA definition requirements.

locates the resource. This command is called a LOCATE. Through the use of this command, a topology database is created to contain the nodes and resources of the network. Thus, topology is built rather than being predefined.

The above method provides advantages over predefinition in ways other than just the time associated with creating definition files. This method also allows for resources to be dynamically found and moved within the network. This enables greater flexibility in both the network design and in backup modes of operation. These located resources can then be added to a topology database on a real-time basis. This database can be used as the foundation for the management of the network.

This database is as accurate as possible because it is created dynamically on a real-time basis. The database contains not only resource names and location, but additional information about the resource. This information includes:

- Node type (EN or NN)
- Session characteristics (from the BIND)
- Window size supported

Figure 5.5 shows how the topology database is built. As information about each of the nodes in the network arrives at the network node, the topology database is updated with the included information.

5.3.3 Definition—HPR

HPR follows the same pattern as APPN. The definition requirements are not changed with the use of HPR. The only distinction is that HPR-capability is included in the topology database.

5.3.4 Definition—TCP/IP

TCP/IP has approximately the same definition requirements as APPN; in TCP/IP, local information consists of the IP address of the node, the subnet mask, the type of driver that will be used on the node, and, possibly, a gateway address. This information is sufficient to allow the node to communicate across an IP network. It also allows a node to initiate a session with a specific node on a specified port.

A TCP/IP network dynamically learns about the nodes that exist in the network. In addition, each node participates in the routing of

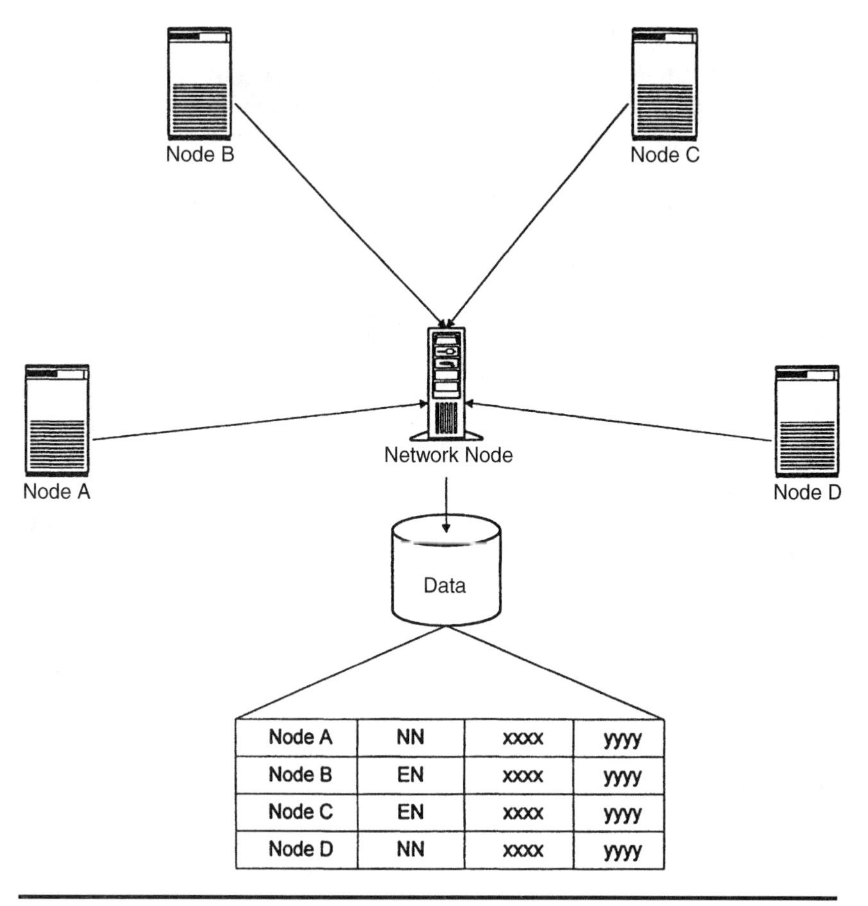

Node A	NN	xxxx	yyyy
Node B	EN	xxxx	yyyy
Node C	EN	xxxx	yyyy
Node D	NN	xxxx	yyyy

Figure 5.5 Topology database build.

messages from their origin to their destination. There is no need to define nodes or to create routing tables. This is done automatically by the network.

5.4 Routing

The four network architectures also embody different philosophies on network routing. These range from predefinition through a dynamic, distributed routing method.

5.4.1 Routing—subarea SNA

Subarea SNA uses preconfigured routing tables that are created by the system programmer. These routing tables are defined in VTAM and NCP and define the physical connectivity between nodes. These tables are called *explicit route* tables. VTAM also defines *virtual route* tables. The virtual route tables define the logical routes that are available for use through the network. The virtual routes are directly mapped upon the explicit routes.

The task of creating the routing tables can become so burdensome and complex, that IBM developed a program that performed this task. This product, Routing Table Generator (RTG), receives as input definitions of the physical connections that exist within the network, some parameters that define the level of redundancy desired, and some other miscellaneous parameters. RTG processes this information and constructs explicit and virtual route tables for each node within the network. Each node has its own set of tables. This fact makes the task of creating the tables that much more burdensome, because it must be reproduced for each routing VTAM and NCP within the network.

The preconfigured routing tables offload the network from the complex, and potentially computationally intensive, routing determination. This enables routing decisions to be made very quickly. Data is passed almost directly from the input queue of a node to the proper output queue. The only decision is whether the desired route is available.

The determination of a route is done by the originator of a session in concert with the establishment of the session.[22] By the source being responsible for route selection, the source of a session is quite important in the network. This is because the actions of the network can be dependent on the actual source of a session. For example, if nodes do not have exactly equivalent routing tables, the route selected can be different as a result of the session being initiated from a different source.

Subarea SNA provides for the potential definition of multiple-output queues.[23] These queues are segregated by a specific route with priorities laid upon those routes. This allows for a great deal of fine-tuning of the routes and scheduling of data to a destination.

---

[22]Subarea SNA does not provide for the use of datagrams between points in the network. All communication is done on a session basis.

[23]Twenty-four output queues can be defined. They are broken down into eight virtual routes between a pair of nodes, with three priority levels within each route.

5.4.1.2 Routing—APPN. APPN dynamically determines routes based on the topology that exists at the start of a session. The difference between APPN and subarea SNA is that APPN does not use preconfigured routing tables, even though routing tables can be preconfigured. Instead, a route is calculated by an NN when a request for a session is received. The NN uses the topology database and preconfigured session parameters. These parameters are used to calculate a route from the session originator to the destination. Although the route is calculated by the NN server for the session, the route does not have to pass through this, or any, NN.

APPN also uses *weights* that are configured for routes to determine which route to use for a session. These weights allow certain routes to be made more attractive to the routing algorithm when calculating a route. Such factors as bandwidth, security, and cost can be included in the route calculation.

APPN routes are calculated once by the network node server of the session originator during session establishment. Although the topology of the network may change over time, these changes are not reflected in the route that is used for a session. Thus, it is possible for a session to be established across a nonoptimal route and the route will not be modified when a more appropriate route becomes available. Figure 5.6 shows an example of this process occurring.

5.4.1.3 Routing—HPR. HPR uses a combination of the dynamic routing of APPN and the source route setup exhibited by token ring networks. With the use of RTP connections, HPR allows for the use of ANR routing.

This routing technique allows for the determination of a source route to a destination. Once the route is chosen and the route setup protocol is completed, a semifixed route is formed across the HPR network. This route forms the basis for the RTP connection.

Many data types can be passed across the RTP connection. Each of these passes along the same path (assuming that the same RTP connection is used). Each intermediate node along the route almost immediately routes packets by using the established route and utilizing the ANR labels within the packet headers. These labels allow each node to quickly determine the next segment along a route and queue the packet to the correct outbound queue.

The use of an RTP connection as the routing basis also provides for the rerouting of data traffic if a segment along the RTP connection

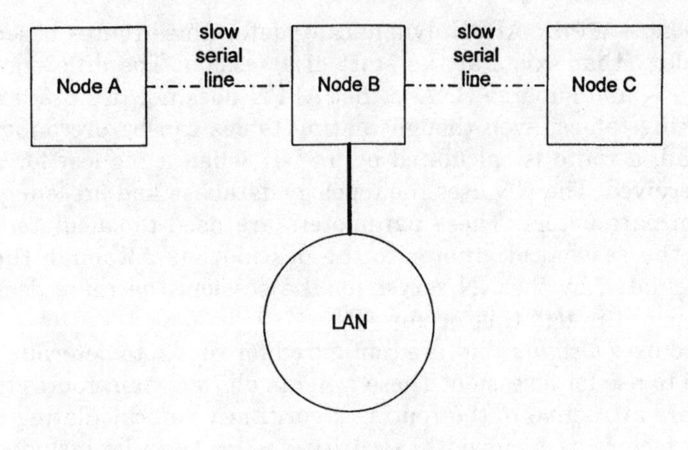

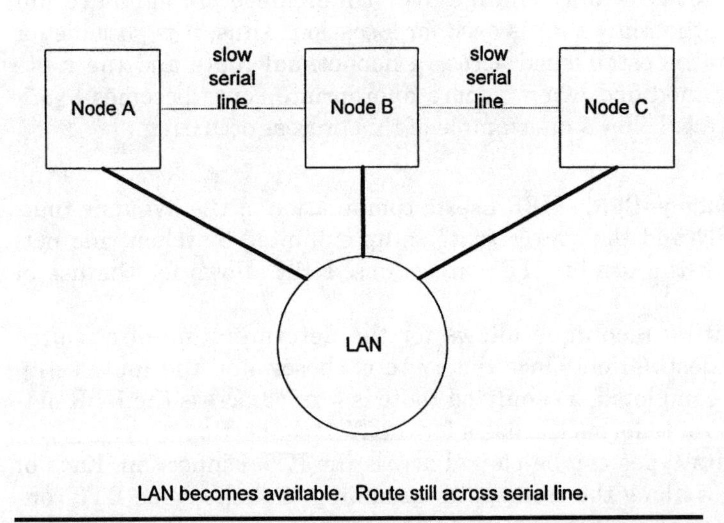

Figure 5.6 Nondynamic route choice.

should fail. If a break in an RTP connection route is detected, the originating node recalculates a new route across the HPR network. Once this route is established, data uses this new route without notifying the LUs on each end of the RTP connection. This provides for this data a transparent routing redundancy not exhibited by subarea

SNA or APPN. The usability profile of the network connection is thus greatly enhanced while not requiring any modification of the endpoint applications (LUs).

5.4.1.4 Routing—TCP/IP. TCP/IP uses a dynamic route scheme. The routing tables are updated through the use of one of the routing protocols, such as RIP or OSPF.

Routes are calculated by each node on a hop-by-hop basis. This allows IP to provide the optimal route at the time of message transmission. In addition, messages are routed by packet. That is, if a message is larger than what the data link layer supports, messages are packetized to allow transmission. This can result in the possible duplication of packets, but as a consequence, can provide connectivity when other protocols would experience a failure.

The routing algorithm that is used can make a large impact on the overhead associated with the routing task. However, this responsibility is not minimal. Routing must be done on a continual basis and can have a negative effect on network throughput.

Routing is a complex task and may be computationally intensive. Because IP must perform this task continually, the overhead associated with routing is higher than the other architectures. But the robustness of message transmission is also higher. These counterbalancing factors must be taken into consideration when comparing IP to the other architectures.

5.5 Integrity

It is generally an absolute requirement that networks provide a high level of integrity for the data traversing through the network. If this integrity is not provided, users will not feel comfortable that the network is providing the correct data. Instead, users will feel that the network is unreliable and will not trust the network to correctly pass their data. For obvious reasons, this is not the objective that is desired in a network.

5.5.1 Integrity—subarea SNA and APPN

Both subarea SNA and APPN embody the point of view that the integrity of data passing through the network is of paramount importance. Both have included extensive detection schemes and recovery procedures to guarantee that data arrives correctly and in order. These steps include the use of network and transport protocols that continually test and correct errors in the data, and to the sequencing of messages that pass through the network. Data integrity of subarea SNA and APPN are one of the major factors that enables them to dominate the networking industry.

Both subarea SNA and APPN also include extensive flow control algorithms. Although there are areas in which the algorithms are not the same,[25] they provide time-tested methods of ensuring that data does not flood a node. This can cause a loss of messages which results in costly retransmissions. The result of these actions is a network that takes on the task of data integrity.

5.5.2 Integrity—HPR

When data is using an RTP connection, the integrity of data is the responsibility of the RTP endpoints. Intermediary nodes along the RTP connection route provide a reduced error recovery procedure. This reduces the overhead associated with an intermediary node, but requires that this responsibility is assumed by someone—the RTP endpoint in this case.

These endpoints provide the error detection and recovery provided by intermediary nodes in subarea SNA and APPN. These endpoint nodes pass indicators representing communication status to ensure that the communication is successful. If an error is detected by one of these nodes, a request for retransmission is passed to the other connection endpoint.

The session endpoints that use the RTP connection for communication have no awareness that there is any difference in communication. These applications/LUs have awareness of the same interface as when subarea SNA or APPN are utilized.

If a segment of an RTP connection should fail, one of the RTP endpoints initiates the creation of a new RTP connection that uses a different set of segments.[26] If the rerouting is successful, the data traversing the new RTP connection will appear the same to nodes

---

[25] APPN includes a patented adaptive flow control algorithm.

[26] This new route must use only HPR nodes. In addition, the endpoints of the HPR network must also support the transport option, so that a new RTP connection can be established.

outside of the HPR network and session partners using this route will be given no awareness of the use of a new path.

5.5.3 Integrity—TCP/IP

TCP/IP takes a different view of network integrity. Rather than seeing this as the task of the network, TCP/IP sees this as the task of a higher-level protocol.[27] Since IP operates only with datagrams, there is no retry or resequencing of messages available from this layer. Instead, these tasks are the responsibility of a higher-level protocol called TCP. The reliance of TCP, at a higher level, for these tasks is not, unto itself, an issue. It can be an issue that these tasks are performed at the ends of the communication path. Thus, messages are retransmitted at endpoints, not at intermediary nodes. This movement of message retransmission to the periphery can result in additional data traffic through the network.

Figure 5.7 illustrates the difference. Retransmission done at the endpoints may result in increased bandwidth utilization. For example, if an error occurs between Node C and Node D, rather than performing a local retransmission between these two nodes, TCP would require a retransmission between Node A and Node F. This would cause the bandwidth to be wasted when transmitting frames from Node A to Node B because this has already been accomplished successfully. If the failed connection is between Node E and Node F (assuming that the data is passing from Node A to Node F), then this retransmission will clog all other nodes along the route.

5.6 Prioritization

When data traverses through the network, it is common for users to feel that "some data is more equal than others." By default, networks look upon all data as having the same characteristics. But in reality, all data is not the same and should not be treated as if it were. Some requests must be responded to within a very short time. Other data can tolerate a lower-priority rating. The network should allow these distinctions to be made and respect them.

---

[27]This is not to say that TCP/IP is unreliable. Rather, the paradigm that IP uses places responsibility on the session layer, not the network or transport layers. This can result in a retransmission of messages at the endpoints. This may be acceptable in a LAN environment, but can be less desirable in a WAN environment.

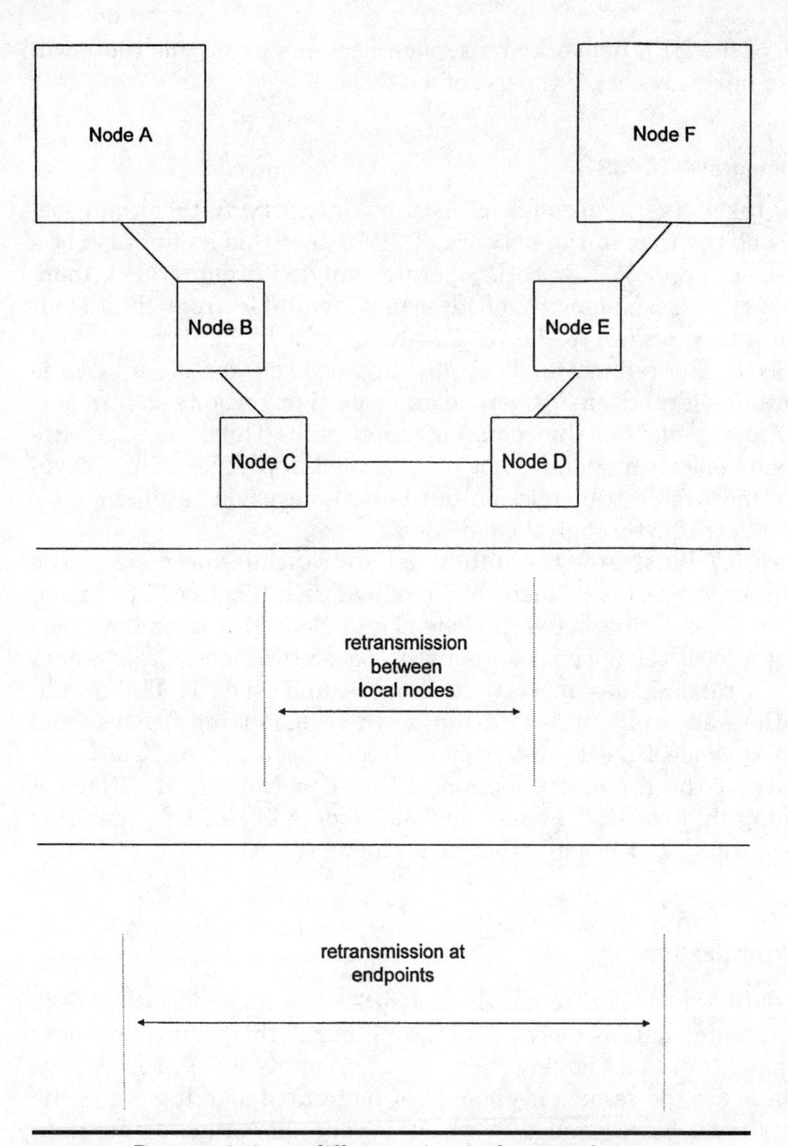

Figure 5.7 Retransmission at different points in the network.

Each of the network architectures dealt with in this book view both the ability to prioritize differently and provide different capabilities for prioritization to be done. Understanding how each network architecture views this important area will assist us in providing an integration method that supports as much functionality of each architecture as possible.

Prioritization is important only when there is less bandwidth available than there is data to be transmitted. If there is adequate bandwidth for all of the data to be transmitted, the ability to specify which data goes first is of minimal interest. When there is more data to be transmitted than bandwidth available, it becomes especially important to specify which data comes first.

5.6.1 Priority—subarea SNA

Subarea SNA provides the tools to design and implement a wide range of prioritization schemes. The architecture allows up to 24 output queues to be specified to each destination domain.[28] This scheme is broken down into eight virtual routes with three transmission priorities within each virtual route. The purpose of these queues is to allow session data to be segregated into queues which are then transmitted to an output port. There is also an additional queue that network control data goes into called the network priority queue.

Priority of the session is specified during session setup through a class of service (COS) table. This table defines the priority of session traffic and specifies which virtual routes to use for the session. The COS table specifies a set of virtual route and priority pairs. This defines the possible routes and priorities that a session can utilize. A COS specifier is added to the session setup parameters.

Table 5.1 shows what a COS table looks like. The COS table consists of sets of virtual route (VR) number/transmission priority pairs. Sessions can be restricted to a specific route by defining a special COS entry that contains a specific VR. If that VR is not available, the session cannot be established.

This can be useful when providing "special" session attributes. These attributes could be something as simple as specifying a certain set of routes or restricting a session to a specific route; if that route is not available, the session cannot be completed.

[28]This number has been expanded in recent releases of VTAM.

TABLE 5.1 Typical COS Table

ISTSDCOS	COSTAB	
ISTVTCOS	COS	VR = ((0,2), (1,2), (2,2), (3,2), (4,2), (5,2), (6,2), (7,2))
INTERACT	COS	VR = ((0,2), (1,2), (2,2), (3,2), (4,2), (5,2), (6,2), (7,2))
NORM	COS	VR = ((0,1), (1,1), (2,1), (3,1), (4,1), (5,1), (6,1), (7,1))
BATCH	COS	VR = ((0,0), (1,0), (2,0), (3,0), (4,0), (5,0), (6,0), (7,0))

Prioritization can also be established by specifying the service order of the polling of downstream devices. This can entail the possible reordering of servicing from its normal sequence. This allows specific devices (PUs) to be serviced more frequently than others.

5.6.2 Priority—APPN and HPR

APPN and HPR provide a process for the COS table that is similar to that of subarea SNA. Because these architectures do not contain virtual routes, which are the basis of the routes in the subarea SNA COS table, the foundation is different.

The APPN network node is the point at which routes are calculated. Because subarea SNA, APPN, and HPR all operate on the basis of source routes and view the resolution of the COS entry only during session startup, the NN becomes the point where a COS mode name is resolved and taken into account.

The NN contains a COS database. This database consists of the following set of fields (Table 5.2):

TABLE 5.2 Contents of APPN COS Table

List of mode entries	Each entry contains a mode name and a corresponding COS entry.
List of COS entries	An entry contains a COS definition, the transmission priority, and the weight assigned to the COS.
Weight index structure	This structure allows actual weights to be calculated once and then stored for future use.

The APPN topology and route selection (TRS) component uses the COS database and the specified mode name during session establishment to select a route for a session. TRS uses the COS table to pick an appropriate route for the session. Although this provides a method of picking routes, it does not provide the same type of reordering of the transmission queue from a node.

APPN COS processing is only operable during session establishment; there are no additional output queues available. Data is transmitted on a first in, first out (FIFO) basis. The idea is that there is excess capacity in the route to provide satisfactory service to all sessions.

In addition, the ISR nodes do not partake in COS processing. This situation is actually similar to the processing done in subarea SNA, where the intermediate routing nodes (VTAM and NCP) are not aware of the sessions that traverse through the network.[29]

There are also some reserved COS entries. These entries are used by specific services within the APPN node. Examples of these entries and the associated services are shown in Table 5.3.

The APPN COS processing is based on specifying characteristics for a TG.[30] The type of TG characteristics are shown in Table 5.4. These characteristics are taken into consideration during route calculation to determine the appropriate route to be used for a session. As is true of subarea SNA, a COS can restrict a session to a specific set of routes. If that route is not available, the session cannot be started.

[29] The session origin and destination are the only nodes that participate in COS considerations. The intermediary routing nodes only have awareness from the point of view of a node on a virtual route.

[30] APPN TGs directly correspond to physical connections.

TABLE 5.3 Transmission Group Characteristics

Property	D(ynamic) S(tatic)
Cost per byte: relative cost of transmitting a byte. Useful when using a public WAN, such as X.25 network. Can also allow users of the network to pay for the creation of the network. This is a user-defined classification.	S
Cost per connect time: applicable when a connection uses a network that charges for connection. A user-defined classification.	S
Security level: indicates the level of security protection provided by the TG. This is an architecturally significant value.	S
Modem class: useful if a specific type of facility is necessary. This might be a specific modem type or a set of modems.	S
Effective capacity: the highest transmission rate before the TG is considered to be overloaded. This is specified as a 1-byte floating point value, expressed in units of 300 bps.	S S
User-defined category 1–3	S
Propagation delay: the time it takes for a byte to reach the other end of the TG. This is expressed in 1-μs intervals.	S or D
Quiescing: specifies that the TG is quiescing. If this value is TRUE, the TG is considered unusable.	D
Operational: a binary value that specifies that the TG is operational and ready to be used for the transmission of data.	D

TABLE 5.4 Comparison of Protocol Headers

Protocol	Header	Default Data Size (bytes)	Percentage* (%)
Subarea SNA	29 (26 + 3) routing nodes	4096	0.7
	9 (6 + 3) peripheral nodes	255	3.5
APPN	9 bytes	255	3.5
TCP/IP	44 bytes	1500	2.9

*The percentages are given for comparison purposes. All of the architectures can support various message sizes.

5.6.3 Priority—TCP/IP

TCP/IP provides very little support for the prioritization of traffic. The intention of TCP/IP is to be a democratic, egalitarian architecture that does not provide better service for a specific user or for certain traffic.

Some implementations of TCP/IP support *priority routing*. This feature is supported by some TCP/IP routers as an option and allows for the specification of a priority to certain traffic. This delineation can be based on protocol, type of service, or some other criteria.

The inherent problem in the use of this specification is that few routers provide this support. Even if the routers do support this option, few endpoint implementations adhere to the use of this option.

5.7 Overhead on the Link

All of these network architectures require the use of some headers, in addition to the actual user data. These headers take the same amount of transmission bandwidth as user data. Thus, the larger the header, the less user data that can be transmitted. Figure 5.8 illustrates how the size of the header effects data throughput.

Table 5.5 shows the header length for each of the protocols. Note that the maximum and default data sizes are also different, so I have added a percentage for the default case.

Table 5.6 compares the network architectures for a constant message size. It is quite evident the differential that is caused by the change in header size. While APPN and peripheral subarea SNA use a 9-byte header that results in a 1.7 percent overhead, TCP/IP uses a 44-byte header (20 bytes for IP and 24 bytes for TCP) that results in an 8.6 percent overhead. This can have a large effect on throughput

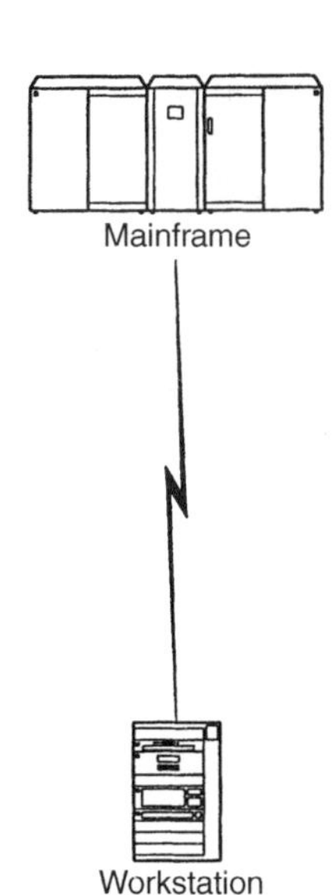

1500 bytes of user data

If the header is 100 bytes and the maximum data size is 200 bytes, then the line efficiency is 50%. In this case, there would be 15 packets required to transmit the data.

If the header is 10 bytes and the maximum data size is 200 bytes, the line efficiency is 95%. In this case, there would be 8 packets. This is a reduction of 47% reduction!

Figure 5.8 Network header size impact.

TABLE 5.5 Comparison of Protocol Headers with the Same Message Size (512 Bytes)

Protocol	Header	Percentage (%)
Subarea SNA	29 (26 + 3) routing nodes	5.6
	9 (6 + 3) peripheral nodes	1.7
APPN	9 bytes	1.7
TCP/IP	44 bytes	8.6

TABLE 5.6 Reserved APPN COS Table Entries

COS Mode Name	Service
CPSVCMG	This mode entry is used for CP-CP sessions. This mode is explicitly supplied when the session is initiated.
SNASVCMG	This mode entry is used for network management.
CPSVRMGR	This mode entry is used for Dependent LU Requester/Servicer (DLUR/S).

when messages of this size or smaller saturate a communications path.[31]

5.8 Summary

The four network architectures provide most of the layers defined in the ISO/OSI network architectural model. The main differential between the SNA architectures and TCP/IP resides in the paradigm of network responsibility in the transport layer. The SNA architectures create a reliable transport network by using data link controls that both detect errors and provide transmission retries, and by providing a similar service in the network layer.

HPR provides the appearance of a reliable transport to those nodes outside of the HPR network, but itself relies upon RTP endpoints to ensure a reliable transport is resultant.

On the other hand, TCP/IP uses a paradigm that the network is not responsible for error detection and recovery. Instead, the endpoints of the network are responsible to ensure data integrity and to provide retransmission logic. This results in a simpler architecture within the network, but with the potential for higher line utilization that results from messages being retransmitted by the endpoints of the network.

There is also a distinction between the session-oriented SNA architectures and the connectionless-oriented IP model. Both SNA architectures rely upon sessions being established between various points of the network. Although subarea SNA has more of these sessions and is more hierarchical in nature, both architectures use only session-oriented connections.

TCP/IP, or more exactly the IP layer, uses only connectionless services. These datagrams require less complexity for the nodes within the network, but higher complexity at the endpoints. This tradeoff is done in an effort to make implementation of the IP stack much easier for vendors to provide.

Prioritization of network traffic is one of the largest distinctions between the two families of design. Both subarea SNA and APPN provide many services that result in a differentiation of priority for different data streams. Among the services are transmission priority, session services use of a COS table, and the modification of the normal polling sequence. The result of these tasks is that network traffic can be segregated into different queues. This provides better servic-

[31]The overhead can become even larger when DLC headers are included. But these headers would be consistent across all of the architectures.

ing for certain network endpoints (while reducing service to others).

TCP/IP contrasts with this view by providing few tuning parameters. TCP/IP expects each node to make an independent assessment of prioritization. If it determines that one type of traffic is overwhelming another, the node can start discarding messages from the latter node. This discarding of data is the way, within TCP/IP, for nodes to reduce their load to acceptable levels.

Neither method is better than the other—they are just different. You must decide which method is appropriate to your situation. One method is not the "correct" one. Instead, they are both available for your use, depending on which fits your requirements. A summary of these four architectures is shown in Table 5.7.

TABLE 5.7 Comparison of Network Architectures

Layer	Subarea SNA	APPN	HPR	TCP/IP
Application	N/A*	N/A	N/A	N/A
Presentation	Some specification. These include 3270, GDS	Some specification. These include 3270, GDS	Some specification. These include 3270, GDS	
Session	Session-based architecture Strong session layer Strong hierarchical typing Network management strongly defined Chaining Sequence number checking per LU Pacing Session rules Bracketing Request types Session types: SSCP-SSCP SSCP-PU SSCP-LU LU-LU	Session-based architecture Strong session layer Peer-peer Network management strongly defined Dynamic LOCATE for destinations Session types: CP-CP LU-LU	Session-layer built on APPN base Sessions flow through connection-oriented RTP connection Route setup protocol builds the ANR labels for the RTP connection	TCP only Sessionlike interface Ports in network Peer-peer Network management outside architecture Error detection at endpoints Datagram base Dynamic routes Little error detection
Transport	Data based on session Relies on error detection at lower layers Prioritization Class of service	Data based on session Relies on error detection at lower layers Prioritization Class of service	RTP connection Transparent pipe across HPR network Error recovery at RTP NCE same prioritization as APPN and subarea SNA COS Rerouting of RTP connection without session impact	
Network	Static routing Predefined routing tables Segmenting Error detection end-end optional Based on error detection at link layer	Dynamic routing Routes calculated at start of session Segmenting Topology updates can affect network throughput Based on error detection at link layer	Source route RTP connection—"virtual pipe" ANR labels Route setup protocol No error recovery at intermediary Low overhead on intermediary nodes	Dynamic routing Routes calculated on hop-by-hop basis Segmenting Error detection only end-end

TABLE 5.7 Comparison of Network Architectures (*Continued*)

Layer	Subarea SNA	APPN	HPR	TCP/IP
Link	Uses link error recovery Not keyed to specific link layer	Uses link error recovery Not keyed to specific link layer	No link error recovery Not keyed to specific link layer	Uses link error recovery Not keyed to specific link layer
Physical	All DLCs supported	All DLCs supported	All DLCs supported	All DLCs supported
Definition	Predefinition	Dynamic topology	Dynamic topology	Dynamic topology
Routing	All routes are preconfigured	Routes are dynamically determined based on existing topology Route session-based Route determined at start of session All data for session follows calculated route	Routes are dynamically determined based on existing topology RTP connection forms a source route RTP connection path can reroute No session endpoint knowledge of new route	Routes are dynamically determined based on routing updates received Route determined on hop-by-hop basis
Integrity	Error-free base created by network and transport layers Capability to generate definite response	Error-free base created by network and transport layers Capability to generate definite response	Error-free base created by network and transport layers Capability to generate definite response	Errors determined by endpoint of data transmission Based on datagram—no recovery in network and transport layers
Prioritization	Class of service table Specify how downstream devices are polled to modify priority Priority is session-based	Class of service table Specify how downstream devices are polled to modify priority Priority is session-based	COS table RTP connection participates in COS processing Uses APPN services if RTP connection cannot be used	No prioritization Priority routing in some routers
Line overhead	Routing node—29 bytes Peripheral—9 bytes	Peripheral—9 bytes	Routing mode—variable based on the number of ANR labels Peripheral—9 bytes	TCP—44 bytes

*Abbreviations: N/A, not applicable.

Chapter

6

Methods of Integration for Subarea SNA and APPN

Although subarea SNA is the predecessor to APPN, the two network architectures cannot be connected directly. However, there is a requirement to provide a migration path between them. The migration path should be as smooth as possible, while providing full functionality of both the subarea SNA and APPN networks.

This is not as simple a requirement as it initially appears. The subarea SNA architecture has a strong definition requirement; all information for the routing and for most connection requests demand that the information be predefined.

APPN is much more flexible in its requirement for system definition. The location of resources is dynamically determined through the use of the LOCATE command. Both the topology and routing components are on a dynamic basis. This dynamic architecture must be interfaced to subarea SNA, which has an extensive definition mandate.

Some of the requirements for this merged system include:

- Each side must use its own definitions.
- There should be as much functionality as possible.
- There should be a clear migration path to obtain full integration.

A common point is needed that both the subarea SNA and APPN nodes can implement. In addition, management of the interface must be as robust as possible, so that the data and the interface can be managed without a major disruption to the existing flows and operation.

There are two basic methods to create this interface. These methods use the LEN functionality of the NCP or operate through the use of a composite node. These two methods will be discussed in this chapter.

6.1 Low-Entry Networking (LEN)

LEN functionality is a simple way for APPN nodes to enter into the subarea SNA network.[1] As was discussed in Chap. 2, a LEN gateway function enables either LEN or APPN nodes to obtain entry into the subarea SNA network. In the case of APPN nodes, this entry enables the entire APPN network to interface to resources in the subarea SNA network without requiring the subarea SNA network to provide any APPN functions.

6.1.1 Design of LEN node

A LEN node was designed as the first step toward a peer-to-peer network. The purpose of the move toward peer-to-peer connectivity was to allow nodes to operate without the intervention of an SSCP. This releases the nodes from relying on a higher node for all connection services. LEN is a migratory step that provides the ability to support a so-called independent LU. Figures 6.5 and 6.6 show the differences between dependent and independent LU sessions. Note: The host in Fig. 6.1 is not involved in the LU session.

An independent LU provides support for the establishment of an LU-LU session without the services of an SSCP. To provide this support, a PU 2.1 base is necessary. Even though the PU 2.1 node is also the basis for APPN, the TRS function is not operable for LEN. Thus, directory and topology services are not available to the LEN node.

Since the LEN node does not provide the extended directory and topology services of a full APPN node, it must contain definitions for the location of a destination LU. These definitions must specify both the location of the destination LU and the route to that destination.

[1]Actual LEN nodes can also gain connectivity by using this interface.

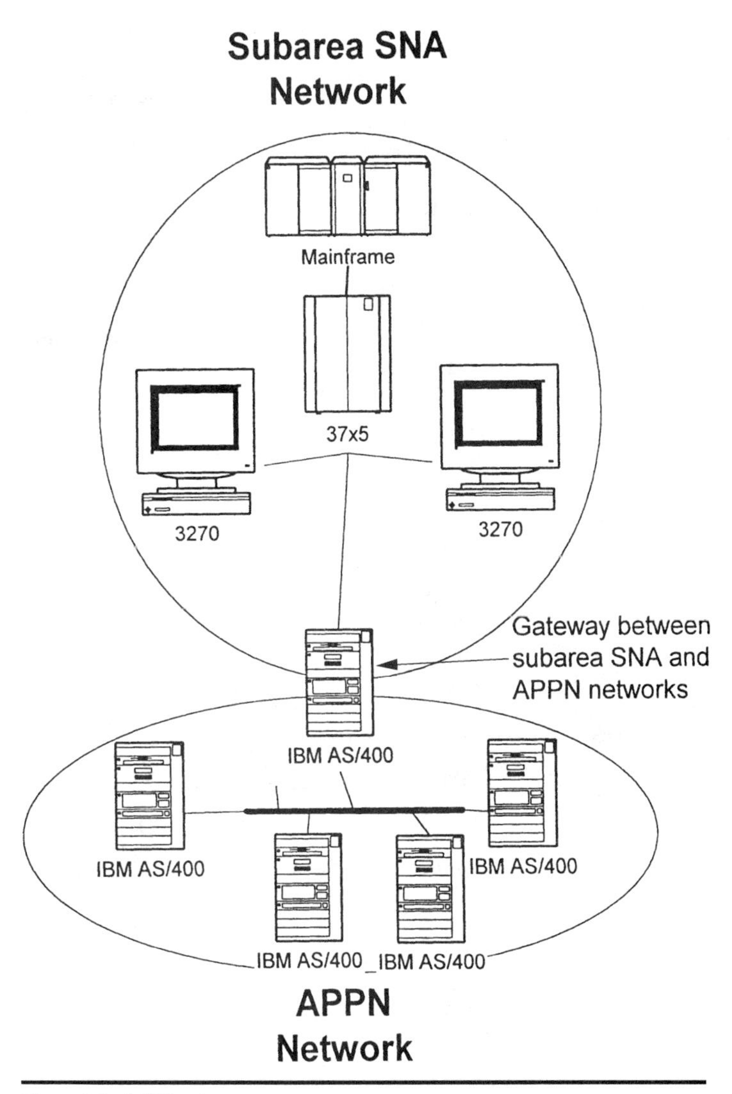

Figure 6.1 LEN gateway.

With this information, an independent LU can transmit a BIND to the partner LU, without requiring directory or topology services. Figure 6.1 shows an example of a LEN gateway.

Figure 6.2 illustrates how definitions are required for operation of a LEN connection. Note that the host has definitions for the LEN gate-

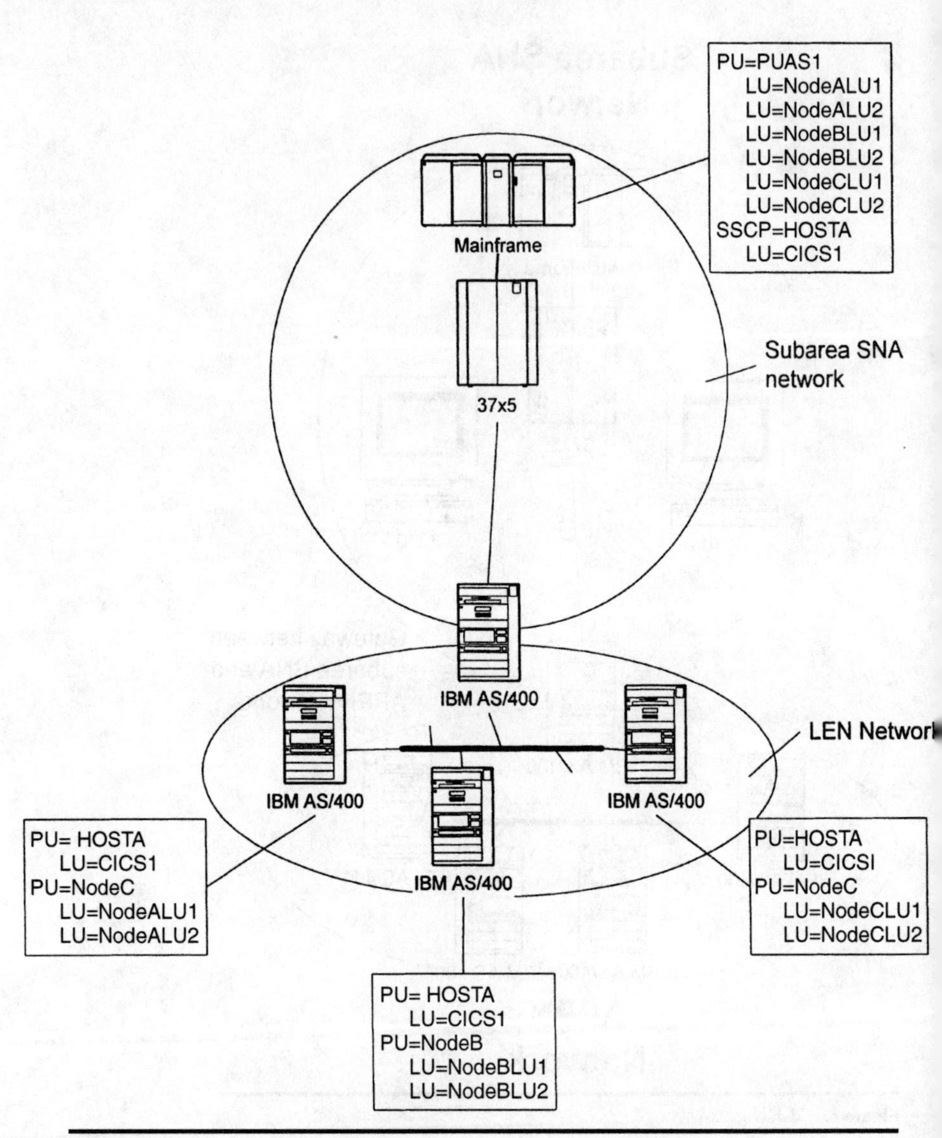

Figure 6.2 Definitions for LEN gateway.

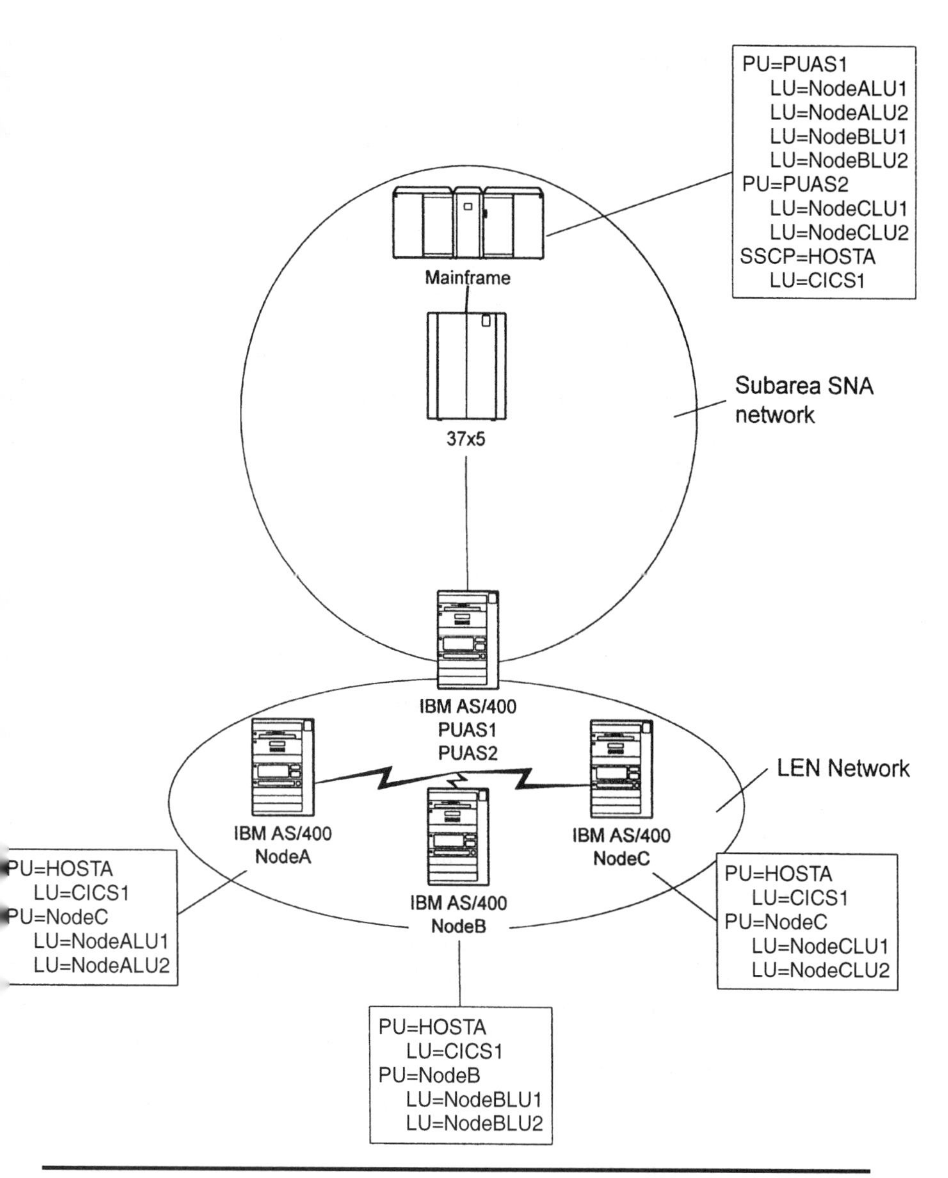

Figure 6.3 Definition for LEN gateway with multiple PU images.

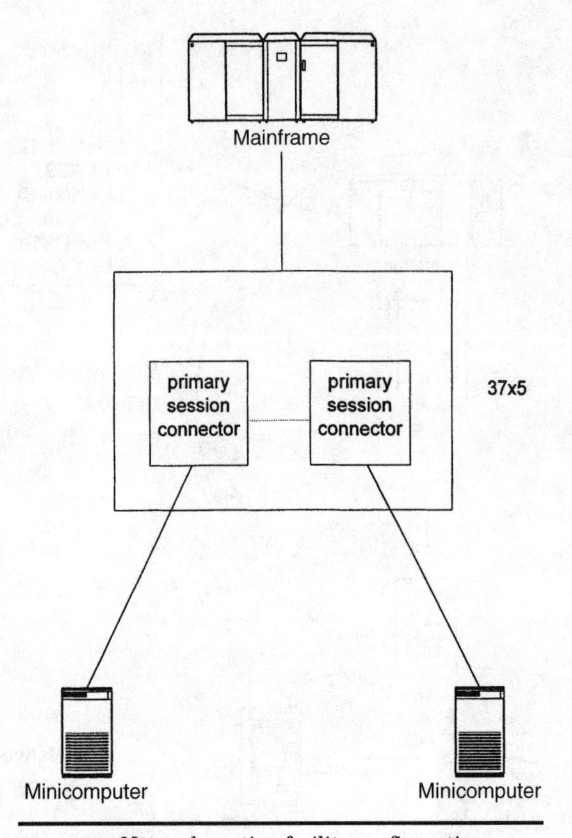

Figure 6.4 Network routing facility configuration.

way and the downstream LUs, but not for the downstream PUs.[2] Instead of appearing connected to their associated PU, the LUs appear to be residing on PUAS1 (the gateway AS/400). Because the LUs appear to be resident on PUAS1, the gateway PU, there is the possibility of exceeding 255 LU limit. This is a design restriction for this configuration.

The definitions in the host and the gateway node allow sessions to be established through the gateway to the host. These sessions are for dependent or independent LUs. The specific method that is used for

[2]These downstream PUs would only appear in NetView through a hierarchical list control vector.

connectivity is dependent on the defined capabilities of the gateway node, the originating PU, and the host.[3]

6.1.2 Limitations

There are several limitations, when using the LEN connection methodology. Among these are:

- A limit on the number of LUs that can interface through a single PU image.
- A directory that is not dynamic.
- Definitions that are required on both the subarea SNA host and the gateway node. Each node in the LEN/APPN network may require a definition.
- Dependent LU support.
- Support for MLTG is not available.

Although this type of connection provides a great deal of service and represents a very viable migration alterative, LEN connectivity has some large restrictions on its functionality. However, most of these limitations are small in scope, they do demonstrate the reasons that APPN was created.

6.1.2.1 LU limitation. A LEN connection uses the services of a PU 2.1 node. This node only uses a FID type 2 for all data transmissions across a network. Because there is only one byte for the relative address (LU address) in this FID, the PU image has an addressing limit of 255 LUs. If this number is exceeded, additional PU images are required to support the additional LUs.

Additional PU images makes managing of the network more complicated, because the PU that a specific LU utilizes must be predefined. The additional PU image must also be represented in the host definitions. This further complicates definition requirements necessary to support this configuration.

It is also necessary to coordinate which LUs are connected to which PUs. For example, Fig. 6.3 shows a network that uses two PU images to provide connectivity to a LEN network. Because definitions are

---

[3]If the host cannot support a "surprise" BIND, then independent LUs cannot gain entry.

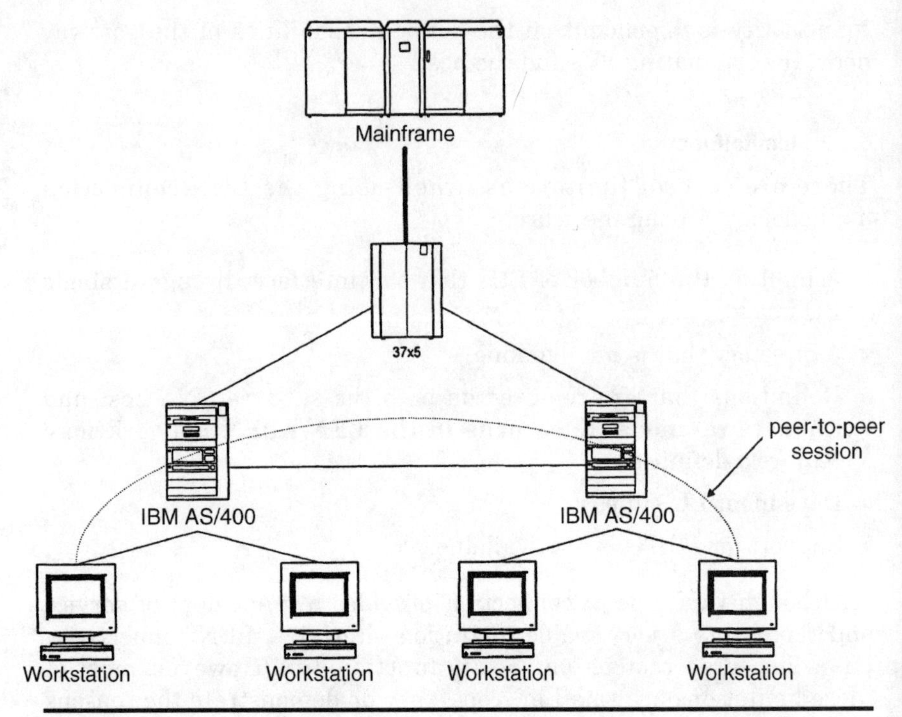

Figure 6.5 Peer-to-peer session.

necessary, if PUAS2 becomes unavailable,[4] NodeCLU1 and NodeCLU2 are unable to establish a connection to the host.[5] In this case, a portion of the network becomes unavailable.

6.1.2.2 Static definitions. The use of a LEN node does not provide any additional capabilities for the support of a dynamic topology. Because LEN nodes are dependent on predefinition as subarea SNA nodes, there is no gain in this area by using LEN connectivity.

The topology information of the network is provided by the definitions that are created. Instead of learning about the network dynamically, as in APPN, LEN nodes have a limited capability to learn the network topology.

---

[4]As the PU image is a software construct, it is possible for one instance to fail while other instances continue to stay available.

[5]There are methods of dynamically adding the necessary definition, but I will not address this issue here.

In addition, the directory of a LEN node consists entirely of the defined components. The directory does not change on a dynamic basis by adding or deleting resource information. If a new resource is added, the LEN definitions will have to be updated to reflect the new resource, if session services are required to that destination.

6.1.2.3 Predefinition. Both the source and destination points of the network require definition when using LEN connectivity. This results in the same definition requirements that exist in subarea SNA nodes.

Differences do exist when moving from subarea SNA to LEN connectivity. The first difference is that the host does not have to be one of the session partners. Subarea SNA has sessions that do not terminate in a host, but these occur only when a component emulates a host. An example of this function is provided by products such as Network Routing Facility (NRF) from IBM. This product enables communication between peer nodes by emulating a host and switching session traffic between two LU-LU sessions. Figure 6.4 shows how NRF provides this connectivity.

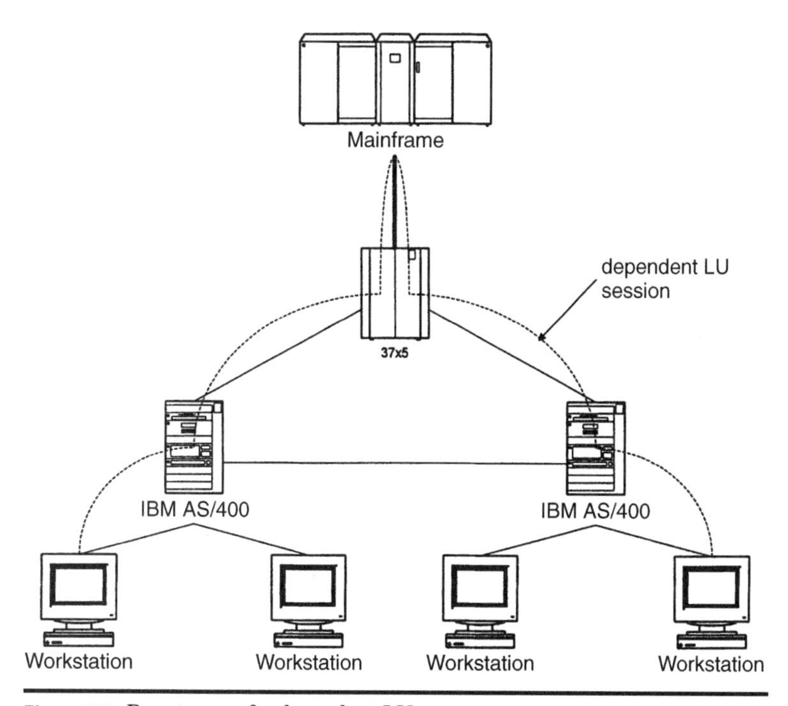

Figure 6.6 Peer-to-peer for dependent LU.

NRF provides primary LU functionality for two sessions to different secondary LUs. This primary feature simulates the functionality of a host application. NRF then establishes a transparent communication path between the two LU-LU sessions. Data from one LU-LU session is directly passed to the other linked session, where it is then transmitted to the other secondary partner. This provides the appearance of having two secondary LUs communicating directly.

The second difference is when independent LUs are utilized. Independent LUs are able to provide session services even when an SSCP is unavailable to provide this function. If the destination of the session is a peer node or a host, a session can be established. This increases the reliability and usability of the network, as connectivity is increased.

Using independent LUs may increase the complexity of network management as a result of the distribution of information across the network. Because information on session status resides in a distributed knowledge base, it is more cumbersome to provide an accurate picture of these sessions than if this knowledge resides in one point, an SSCP, as in subarea SNA. Because an SSCP is not required for session services of independent LUs, the SSCP loses some information on session status. Although SESSST and SESSEND commands may still be sent to the SSCP to delineate a session start and end, respectively, this is not a requirement. Thus, if these commands are not sent, the SSCP loses awareness of the peer session establishment and destruction.

6.1.2.4 Dependent LU support. Support for the LEN network provides a great deal of flexibility in the operation of the network. Resources can be contacted and sessions can be established with resources in the LEN network. Unfortunately, dependent LUs within the LEN network cannot operate unless an SSCP-function is provided because this is normally not available within the LEN network. Thus, a migration path for these resources must be created.

There are basically two methods of providing this functionality. The first method allows the dependent LU (and its associated PU) to be "seen" from the subarea network. The normal SSCP in the subarea network is thereby enabled to support the dependent LUs in the LEN network.

The second method provides a simulated SSCP gateway within the LEN network. This method is illustrated in Fig. 6.6. The simulated SSCP within the gateway node[6] allows the downstream dependent

---

[6]This function does not have to be located at the gateway to the subarea network. This is shown for illustration purposes only.

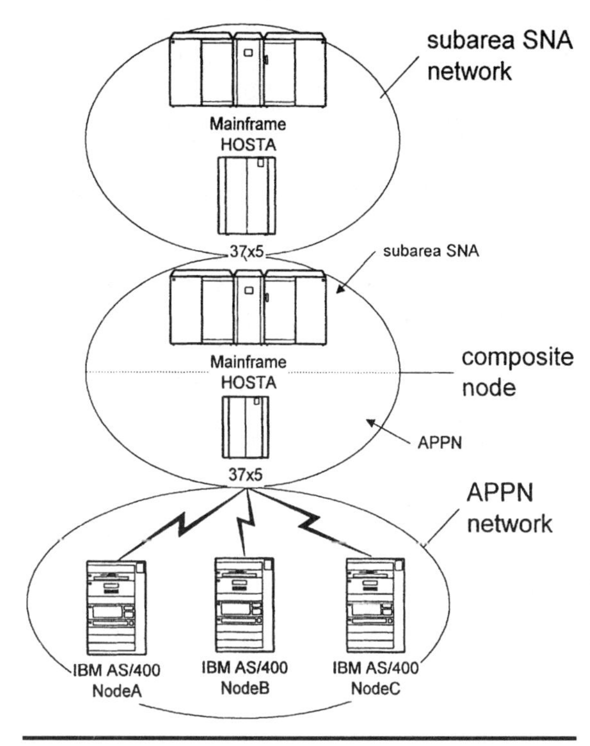

Figure 6.7 Composite node configuration.

resources to be fully supported. The simulated SSCP then passes the user data across an LU0, LU2, or LU6.2 session into the subarea network. This method provides full functionality for the dependent LUs while also allowing easy access into the subarea network.

6.1.2.5 No support for multilink transmission group (MLTG). Although the LEN node may represent an entry into an entire network of resources, because the LEN node resides as a peripheral node, there is no support for MLTG. This results in the removal of MLTG advantages.

For example, if bulk data transfer is occurring into the LEN/APPN network, a single physical circuit is used to transfer the data into the network. Because the session is directly tied to a specific TG, the flexibility of an MLTG cannot be utilized.

As was described in Chap. 1, the MLTG provides many options for connectivity. It provides the ability to represent multiple physical lines as a single transmission group that logically traverses the network. The MLTG also allows lines to be dynamically added and deleted to the group by allowing a great deal of flexibility in the network configuration. This dynamic ability allows the physical network allocation to be modified as required. Thus, lines could be added during a period of bulk data transfer and removed when demand is required for other uses.

Figure 6.8 shows an MLTG between two host systems. If the MLTG were tied to a LEN node, this connection could not be used, unless a single static link is used. However, this method can greatly restrict your options during connectivity. For example, if the same type of periodic bulk data transfer is required, the link to the node would have to be sufficient to carry the data. The rest of the time, this link may have excessive capacity that would have to be paid for when it was not being used.

6.2 Composite Node

Another alternative for connectivity to an APPN node is for the host to provide a gateway facility between subarea SNA and APPN. This method provides a true gateway between the two network architectures.

The gateway function enables the host (VTAM) and the communication controller (NCP) to share the responsibility of creating APPN node functions. At the same time, normal subarea SNA functions are still available and operative.

VTAM provides a bridge between the two network architectures that enables sessions and session data to move freely between the two architectures. This method of allowing sessions to move in and out of either of the architectures was developed by IBM. The APPN topology and route services obtains enough information on the topology to calculate as nearly an optimal route as possible.

A great deal of care was taken to ensure compatibility with most of the functions for both architectures. Thus, a session originating in the subarea SNA side of the composite node may traverse into the APPN network. The knowledge of the APPN network to the subarea side is limited to the destination LU residing on the gateway.[7] Although

[7]The same case is discussed when reviewing the LEN configuration.

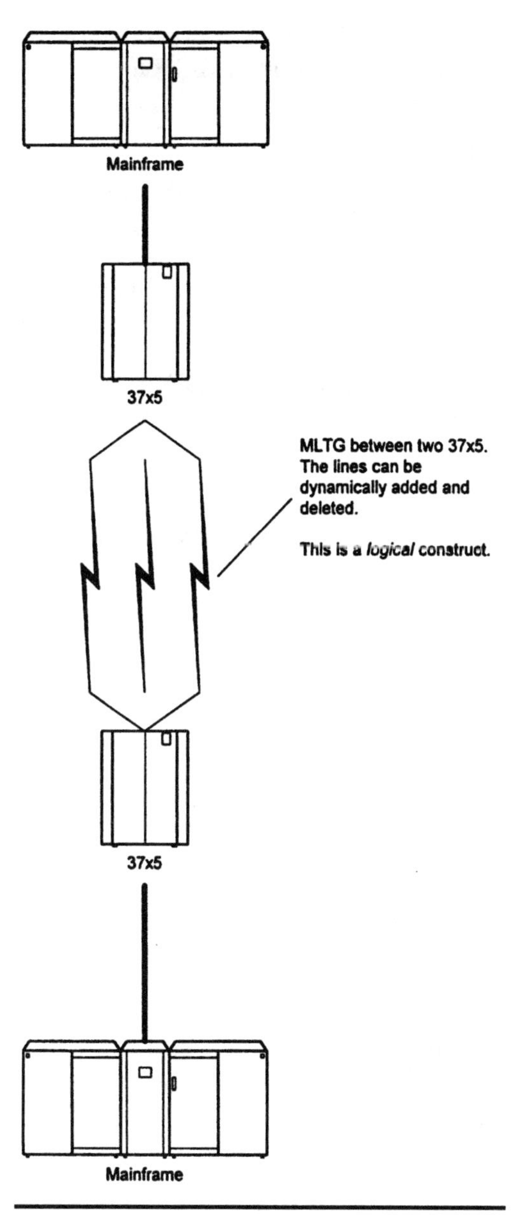

Figure 6.8 Multilink transmission group (MLTG).

routes may not be optimal in getting to the gateway, once entering the APPN network, normal route selection services are available.

6.2.1 Design of the composite node

A composite node provides a dual image for each network operation and service. This dual image allows each network architecture to provide a full set of features. In essence, this node type can be thought of as the emulation of two full networks. Figure 6.7 shows this configuration.

This figure also illustrates that the composite node acts as a single entity because the operational distinction between VTAM and NCP does not exist. This fact becomes important as we investigate this node type and its operation.

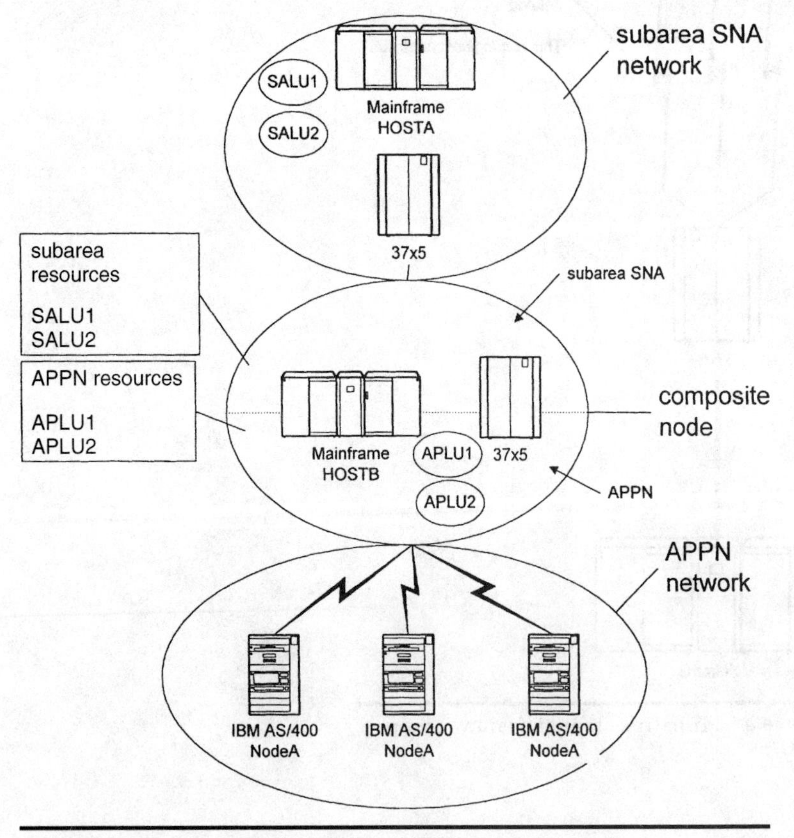

Figure 6.9 Composite node knowledge of APPN network.

The composite node is formed by VTAM and the NCP in a synergistic relationship. The tightly coupled system provides a single image to the APPN network of one node. This composite node can either provide an EN or NN functionality. To the subarea SNA network, VTAM and NCP are seen as they currently are.

This "dual personality" is augmented by the ability of resources in either network to access resources across the architectural boundary. The composite node provides the gateway function between the networks. Both networks gain awareness of each other and are thus able to connect to resources freely.

6.2.2 View from the subarea SNA side of the composite node

VTAM and the NCP have full awareness of the subarea SNA network, but their functions are fully supported by the composite node. Normal processing is available for resources on this side of the network.

The resources on the APPN side belong to a cross-domain resource manager (CDRM). By being defined in this manner, a greater dynamic association is possible. For example, APPN resources can dynamically establish sessions with subarea SNA resources without predefining the APPN resources. This is done by using existing cross-domain logic for unknown resources. These resources are dynamically created by VTAM upon arriving in the subarea SNA domain. Although this is not the same dynamic process that exists in APPN, this was the most flexible method available to the VTAM developers. This method also aids VTAM customers that want to migrate, or at least gain connection, to an APPN network.

By using this method, an APPN network can gain connectivity to the subarea SNA network without having to use a LEN gateway. This allows full directory and topology services to be utilized in the APPN network. In addition, because directory services are extended into the subarea SNA network, resources in the subarea SNA network become known to the APPN network either through the NN function in the composite node or through LU registration.

The subarea resources also gain connectivity to the APPN network and its resources. Because the subarea SNA network requires predefinition of resources to which sessions are initiated by the subarea network, all resources in the APPN network to which sessions are initiated from the subarea SNA side must be predefined. These CDRSCs (cross-domain resources) are defined as being controlled by the CDRM representing the APPN side of the composite node.

Although the resource is defined as residing in the APPN network, the actual location is not defined. Instead, once the request crosses the network boundary,[8] full APPN functionality allows the location of the resource to be dynamically determined. Normal APPN data flows originate from the composite node to locate the resource.

6.2.3 View from the APPN side of the composite node

The APPN side of the composite node sees an APPN network. The SSCP does not reside in this network view. However, normal APPN nodes of ENs and NNs are recognized. These nodes establish normal APPN CP-CP sessions to create the fabric of the network.

The APPN-composite node exists as another APPN node. It can provide either EN or NN functionality to the rest of the network. The joint VTAM/NCP APPN-composite node meets all of the requirements of an APPN node. It establishes CP-CP sessions and supports independent LU sessions originating and destined for LUs under its domain. Thus, the APPN-composite node is a full APPN node.

The subarea-side of the composite node appears to the APPN network as an extension of the APPN-side. Thus, LUs that exist on the subarea-side appear to exist on the APPN-side of the composite node. This is provided through the session services extensions (SSE) of VTAM.

The LUs on the subarea-side are registered on the APPN-side. At that point, these resources are available for sessions from LUs in the APPN network. Sessions destined for subarea resources obtain knowledge of their location from the APPN/composite node. Figure 6.9 shows this process.

The APPN LOCATE command arrives at the composite node. If the LUs are not registered, the request is forwarded to the APPN network, but will also enter the subarea SNA network. Once there, attempts will be made to locate the resource. If it is found within one of the subarea domains, a positive response is returned to the origin of the LOCATE command, along with tail vectors describing how to reach the resource. These vectors describe the resource as residing adjacent to the composite node. Thus, the subarea SNA network is collapsed to the composite node; all intermediary nodes in the subarea SNA network are not seen in the tail vectors.

---

[8] This is not a cross-network connection. I use this term to signify that the session crosses the boundary between the two network types. Both cross-network and SNA network interconnection (SNI) topologies will be discussed later.

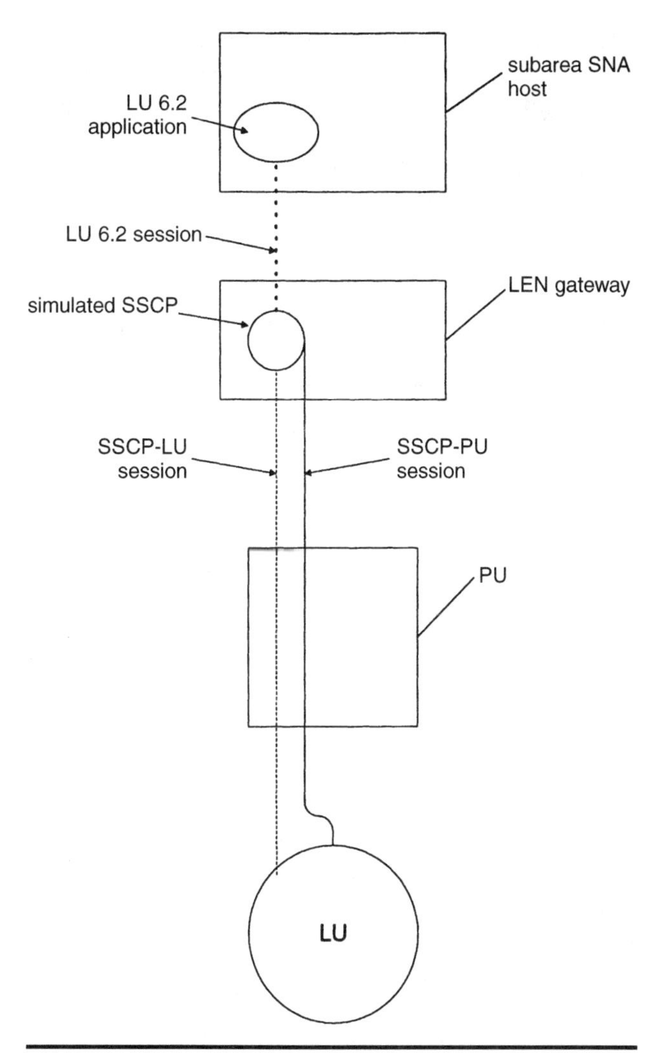

Figure 6.10 Simulated SSCP in LEN gateway.

Session requests reach the composite node destined for the target LU. At the composite node, tail vectors on the BIND are modified to represent the remainder of the path to the resource. On return, the entire path is available, including the network name of the subarea SNA network. This name can be used by the NN of the originator to calculate the best route to the destination.

Network management is handled by using these extended tail vectors. These vectors describe the actual path for the session, including the path through the subarea SNA network. A network management application, such as NetView, can interrogate these extended vectors and determine the exact path for the session, as well as the physical structure of the networks. This information can be used by the network management application to isolate a network failure and extrapolate the failed physical connection.

6.2.4 Limitations

The limitations of using the APPN-composite node are similar to those of the LEN connection. The method I will use to look at the limitations of the APPN-composite node is to contrast the LEN limitations with those of the APPN/composite node.

The LEN limitations include:

- Limit on the number of LUs that can interface through a single PU image
- Directory is not dynamic
- Definitions are required on both the subarea SNA host and the gateway node, plus possible definition on each node in the LEN/APPN network
- Dependent LU support
- Support for MLTG is not available

6.2.4.1 LU limitation. The LEN has a limit on the number of LUs that can be configured to use a single PU image to the LEN network. The FID type 2 that is used for this connection is the same FID that is used for the APPN connectivity.

The difference is that the local address used in the FID type 2 is session-based, rather than fixed for an LU. This local identifier is called a local form session identifier (LFSID) and is generated when a session becomes established. The address space manager (ASM) generates the local address for each session, while the ISR nodes generate their own number. This results in a session being related to different local addresses along the path of a session.

The LFSID is a 17-bit identifier that consists of the following components:

- 1 bit for the ODAI (OAF-DAF Assignor Indicator)
- 8-bit SIDH (session identifier high)
- 8-bit SIDL (session identifier low)

The ODAI is set by the primary link station as 0 and 1 when set by the secondary link station. This divides the address space into two distinct domains and prohibits the primary and secondary link stations from picking the same LFSID.

The SIDH and SIDL provide for an address space containing 65,536 unique identifiers per TG. This space is further divided into the following ranges (Table 6.1):

TABLE 6.1 SIDH and SIDL Values Range

SIDH-SIDL Range	Significance
X′FF00′–X′FFFF′	Reserved
X′0200′–X′FEFF′	CP-CP and independent LU-LU sessions
X′0101′–X′01FF′	Dependent LU-LU sessions or, if dependent LU-LU sessions are not supported, CP-CP and LU-LU sessions
X′0100′	BIND flow control
X′0001′–X′00FF′	SSCP-LU sessions or not at all
X′0000′	SSCP-PU sessions or not at all

This partitioning of the address space enables PU 2.1 nodes that support dependent LUs, to use the existing address range, while expanding the address space used by independent LUs.[9]

Because the LFSID is generated on a session basis, as they are taken down, those LFSIDs are returned to the available pool of numbers. Thus, the number 65,536 reflects the number of concurrent sessions.

This expansion of the addressing scheme enables the APPN connection to support a larger number of services. Although it is still possible to exhaust the number of LFSIDs, it is not feasible to provide connectivity to 65,536 sessions across a single TG. Thus, the LU limitation per TG is not a restriction.

6.2.4.2 Static definitions. LEN connections are defined statically for entry into the subarea SNA network through a specific interface. The APPN-composite node has the same requirement for entry into the subarea SNA network; this is a restriction on the subarea SNA-side of the network. But the APPN-composite node also fully supports dynamic definition, as displayed by any APPN node.

---

[9]The address space supports the existing addresses used in the subarea SNA network.

The location of resources and routes between nodes are fully flexible for APPN-composite nodes. These nodes provide full support for the dynamic movement of resources. Logical unit registration by both the APPN and subarea SNA-sides allows connectivity to all resources available on the node and in both networks. The composite node dynamically moves definitions and enables all APPN services to recognize this movement. Routes are calculated to reflect the network topology at the time of session establishment.

6.2.4.3 Predefinition. The LEN node requires predefinition before establishing a session. The APPN-composite node does not have the same requirement. Although sessions from the subarea SNA side of this node require predefinition before establishing sessions, there is no requirement for predefinition for any other session. Sessions destined for the subarea-side are not predefined. Instead, the originator of the session is dynamically created through the CDRM representing the APPN-composite node.[10]

Because static definitions are eliminated for all other sessions, the APPN-composite node gains most of the advantages of an APPN node. In addition, this node allows a gateway into the abundant subarea SNA network arena.

6.2.4.4 Dependent LU support. The APPN network only supports LU 6.2 sessions for independent LUs. An example of this type of session is the CP-CP session between APPN nodes. Yet there are tens of millions of dependent LUs that must be supported when migrating to APPN. The billions of dollars that these resources represent cannot be thrown away for the sake of migrating to a new network architecture.

A new architecture was developed to aid in the migration of dependent LUs to an APPN network. This architecture, an outgrowth of the APPN Implementors Workshop (AIW), is called dependent LU requester/server (DLUr/s).

The DLUr/s architecture is similar to the dependent LU gateway discussed in Sec. 6.1.2.4. This architecture is based on the idea of a dependent LU requester that operates in the network. The LU requestor supports the establishment of a local session to the remote dependent LU. This object provides local receiver ready (RR) spoofing, while passing all other system requests on to the dependent LU server.

[10]This dynamic operation is exactly how subarea SNA currently operates.

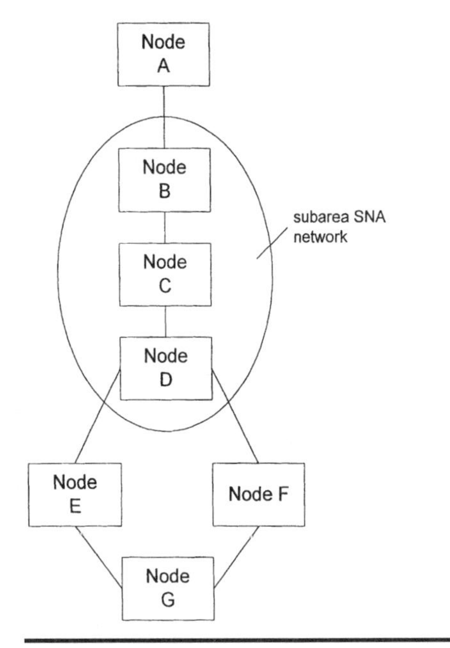

Figure 6.11 Actual subarea SNA and APPN network.

The server is located adjacent to the SSCP.[11] System requests arrive from the DLUr and are passed across a private interface to the SSCP, which operates as if the dependent resource is local. There is no awareness that the dependent resource is actually distributed in the network and that the connection to this resource is disjointed. This view can be seen in Fig. 6.14.

Two independent LU 6.2 sessions support the DLUr/s architecture. These sessions use a new special logmode, CPSVRMGR. The CPSVRMGR sessions can be established by either the server or requester. The only distinction between the server and requester is the definitions that must be available. By allowing the CPSVRMGR sessions to be established from either side, network management is more flexible and provides a clean ability to set up sessions from any available station, no matter what the sequence of activations or inactivations.

In this mode of operation, the DLUr or DLUs can request that a connection be established to an end device. Because the session

[11]The SSCP is normally VTAM. VTAM V4R2 supports DLUR/S architecture.

between the DLUs and DLUr is an independent LU 6.2 session, normal APPN topology and directory services are utilized during session establishment. Thus, the location of the end device can be migrated without effecting the ability of the SSCP in the composite node to find it by the request being transported through the CPSVRMGR sessions. The normal APPN LOCATE commands flow through the special control session between the DLUr and DLUs. APPN directory services is available at each node to locate the LU.

If the resource is on the subarea SNA-side of the network, the DLUs provides the ability to pass the request to the subarea SNA network. A search of the subarea SNA network is conducted to locate the resource. Once located, the DLUs can pass requests to and from the destination.

It is possible in this environment to dynamically register dependent LUs. If VTAM uses SDDLU support, resources can be dynamically allocated. To support this, the DLUr must send an NMVT with a product set identifier (PSID) from the LU. This step must be taken for each dynamically allocated LU. By using this method, not only can dependent LUs ride on the dynamic topology of APPN, but they themselves do not have to be predefined. Thus, the dependent LU gains almost all of the advantages of APPN without any modifications or enhancements.

This methodology provides the subarea SNA network many of the advantages of APPN while not requiring the removal of old equipment. This migration path is smooth and almost all requirements for connectivity are met. The ability to pass data between the subarea SNA and APPN networks for dependent LUs enables APPN advances to be utilized by the "old, antiquated" equipment. Because this "old" equipment represents an order of magnitude larger market size than APPN resources, this migration path is greatly appreciated.

System requests and responses that support the dependent LU are encapsulated across the session between the DLUr and DLUs. Within this session, the entire FID2 SSCP-PU and SSCP-LU traffic is encapsulated. This encapsulated data includes the transmission and request/response headers. The DLUr and DLUs extract the encapsulated data and make any necessary changes before it is transmitted. These changes include creating, modifying, or removing attached control vectors. These manipulations are done to allow as varied a combination of downstream support as possible.

As the CPSVRMGR sessions are used for requests and responses to multiple PUs, there needs to be an addressing method. The fully qualified procedure correlator identifier (FQPCID) is carried in each message and allows unique addressing of the network component so that correct state machines and addressing can be performed.

The FQPCID is created by the node creating the first encapsulated flow. Depending on the direction of initiation, the encapsulated flow is either the sender or receiver. The use of the FQPCID allows a single CPSVRMGR session to address up to 255 PUs.

6.2.4.5 No support for multilink transmission group (MLTG). The APPN-composite node still provides connectivity through a peripheral connection. The same limitations on FID type 2 interfaces still exists on this node. Thus, support for MLTGs for APPN connections is not available.

What is interesting about this restriction is that it does not apply to the subarea SNA-side of the composite node. Thus, the subarea SNA-side of the node still has the capability to use MLTG to connect to another subarea node.[12] If you were to define a connection to another composite node, using the APPN-side for the connection would preclude use of a MLTG, while using the subarea SNA-side would provide this support. Therefore, it is advantageous to use the subarea SNA-side for connections of this type, even though you may feel that this limits your connection options.

Although there is a requirement for more definition when using this method, the support for MLTG may overshadow this small investment in definition. As the definitions would be only for the physical connection and not for the resources at the destination, the additional definitions would be restricted to the routing tables. Though routing table definition is not a small investment, the advantages may overcome this investment.

Note: VTAM Version 4 Release 2 (V4R2) supports a new facility called VR TG. This version provides the image of a TG to the APPN topology database that in reality represents the virtual route of the subarea SNA network. By representing a VR in this manner, the TG is added to the APPN topology, while not requiring changes to the architecture of the boundary connection used by all APPN nodes.

6.3 Network Management

When the view of the subarea SNA network collapses, network management may become difficult. This is because the actual network

---

[12]Either a VTAM host or NCP node.

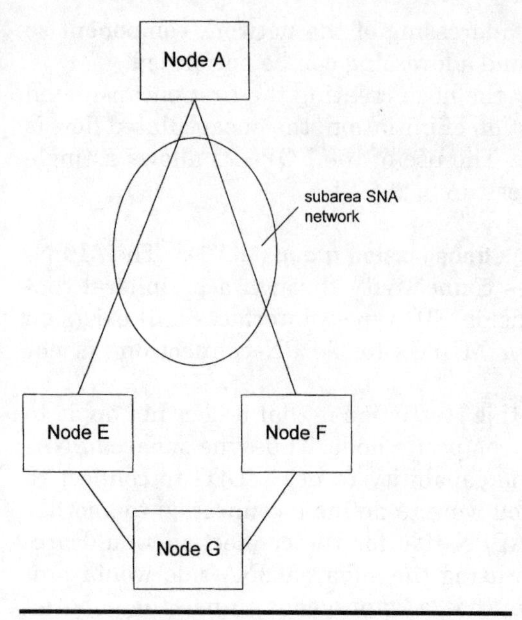

Figure 6.12 Collapsed subarea SNA network.

topology may no longer be seen. The loss of specific topology can aggravate the problem of network management.

For example, the subarea SNA network topology, as seen in Fig. 6.11, collapses from the APPN point of view so that a problem in the subarea SNA network becomes difficult to isolate. The collapsed network, as seen in Fig. 6.12, has no awareness of the subarea SNA network. All that is known is that Node A is connected to Node E and Node F; the intervening nodes are not represented in the APPN network. Thus, network management can become difficult.

One way of closing this knowledge gap is through the hierarchical list control vector that may be within the network management information. This vector provides a detailed description of every node and TG that a session traverses. By analyzing this control vector, a management application can accurately describe the network topology. In referencing Fig. 6.12 and assuming that the only TG is TG1, this control vector would appear as follows for a session from Node A to Node G (Fig. 6.13).

Even though the APPN network has no awareness of the subarea portion of this session path (Node B–TG1–Node C–TG1–Node D), the management application can parse this control vector and depict the actual session path. This method greatly alleviates the problem associated with the network.

Node A-TG1-Node B-TG1-Node C-TG1-Node D-TG1-Node E-TG1-Node G

Figure 6.13 TG vectors.

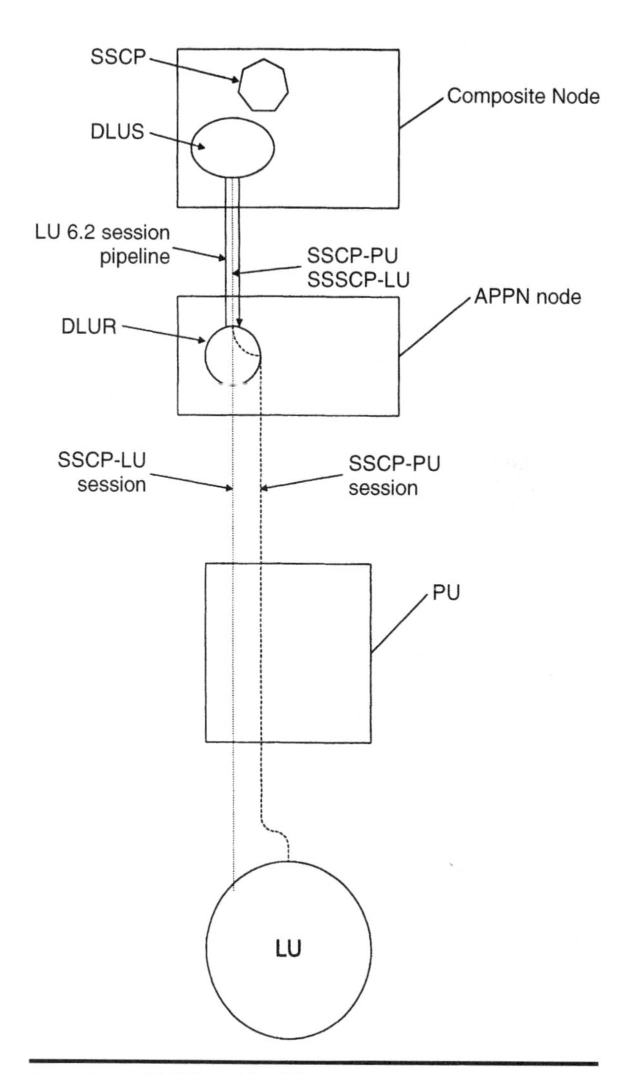

Figure 6.14 DLU/r and DLU/s.

6.4 Cross-Network Connections

The address space for subarea SNA networks is segregated into domains, which cannot contain duplicated resource names. In addition, the network address is separated into subarea and element components.

These networks also provide connectivity across network boundaries. These network boundaries may be isolated within a single company's network, or may be connected to another SNA network. This situation becomes useful when companies are connecting disjointed networks as a result of a corporate consolidation. In this case, the network name is used to determine that the connection is crossing a network boundary.

When crossing a network boundary, it is possible to have little or no knowledge of the physical topology of the other network. It is even possible to have resources with the exact same name as in the local network. In this case, an alias is given to the duplicated resource. This alias allows the network to determine which resource is being specified.[13]

APPN is based on the paradigm of a global address space. The key is that the APPN network node has a full view of the network topology. This may not be possible when crossing a network boundary.

APPN has defined a *border node* function that is similar to that of the composite node. The border node function allows access across the network boundary, between subarea SNA and APPN, except that, in this case, a resource is located across a logical network boundary.

For this configuration to operate it requires a method for allowing LOCATE requests to flow across a network and for sessions to cross the boundary with minimal impact. These requirements have been provided in the APPN architecture for a border node, but the fine details of border node operation has not been completely defined. A complication thus results when supporting this configuration.

In order to support cross network communication, there are two types of border nodes. The first is the *peripheral border node.* This node type provides EN connectivity to ENs across a nonnative network boundary.[14] Because of restrictions on EN functionality, in this case nodes using peripheral boundary node functionality must either be adjacent or must be no more than one hop from a common net-

[13]NetView provides the alias name mapping. Thus, NetView is required to provide this function.

[14]The nodes have different NETIDs.

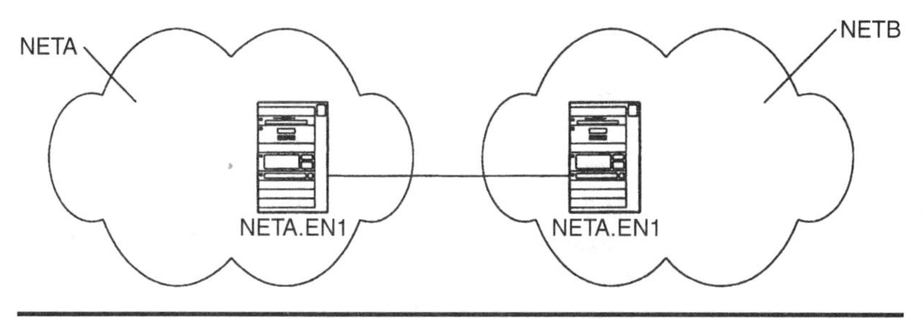

Figure 6.15 Nonnative EN-EN connectivity.

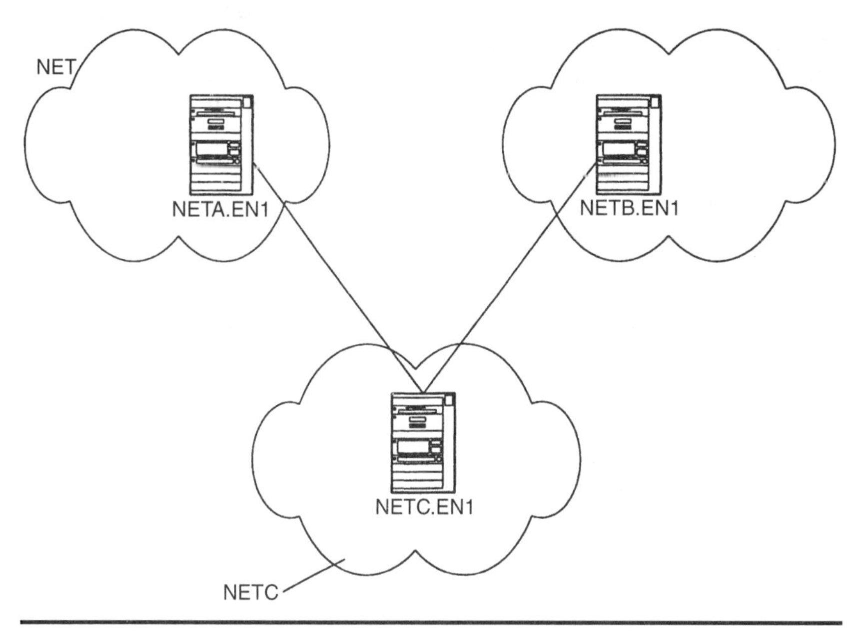

Figure 6.16 Nonnative EN-EN connectivity through an intermediate network.

work. Figures 6.15 and 6.16 illustrate the type of networks that are supported for peripheral boundary nodes.

All APPN nodes support connectivity to an EN that has a different network identifier (NETID) than itself. This configuration exists in both of these figures. In Fig. 6.15, the nodes are directly adjacent. When NETA.EN1 and NETB.EN1 establish CP-CP sessions, their capabilities vectors identify themselves as being ENs, and the NETIDs are different. Because these nodes are adjacent, they are able to communicate. In Fig. 6.16, NETA.EN1 and NETB.EN1 are not adjacent to each other, but they are adjacent to a common network. Because there is no need for an ISR node, these nodes are able to communicate.[15]

The other mode of operation for nonnative network attachment is through the *extended border node.* The difference between the peripheral border node and extended border node is that the latter is able to provide ISR functionality across nonnative connections. These nodes implement the APPN option set 1013. When CP-CP sessions are established between network nodes, the Session Services (SS) component of the node determines if the nodes have different NETIDs and, if they do, it is specified that at least one of them must have an extended border node specified in the CP vector. If the NETIDs are different and neither has this capability, SS deactivates the CP sessions with sense code X′08910006′ (invalid network identifier).

Because many subarea SNA users utilize some cross-network connections,[16] these types of connections pose a serious concern to customers using them. There are very few implementations that support extended border node capabilities. Thus, migration would require nodes that wish to establish a session either be adjacent or share a common node that is adjacent. This does not allow many of the configurations that exist to be smoothly migrated, because these networks do not meet one of the criteria. A great deal of network redesign results. Since one of the major reasons to migrate to APPN is ease in the maintenance of the network configuration, network redesign would not be desired.[17] In addition, this restriction could severely restrict the possible configurations that could be supported.

6.5 Summary

Although APPN is a follow-on to subarea SNA, the differences are extensive enough to make the migration to APPN a challenge. Migration from subarea SNA to APPN can be accomplished through one of two methods.

The first method is to provide a LEN gateway from the APPN network to the subarea SNA network. This method utilizes the ability of

[15]If there were another network between them, then communication would be impossible because an ISR component would be necessary to establish a session.

[16]Often these cross-network connections are clusters within a single company. This is done to isolate networks for different purposes.

[17]Though it may be tolerated if it only had to be done once.

VTAM Version 3 and NCP Version 4 to support sessions to a LEN node. This configuration provides for the creation of a gateway from the APPN network to the subarea SNA network through a single node.[18]

This configuration provides for sessions to be established in either direction through the gateway node. At this point, the gateway node becomes a single point-of-failure; if this node fails, all communication between the subarea SNA and APPN networks also fails. The only way to recover from this failure is to enable a different gateway node and reestablish all of the sessions that were active at the time of the failure.

When using this configuration, there are limits on connectivity between the two network architectures. For example, a connection to a single PU on the LEN gateway can only support 255 LUs in the APPN network. This can become a significant limitation if the APPN network is large. There is also a requirement for full definition within the subarea SNA network. If a resource is not predefined, it cannot be activated or communicated with from the subarea SNA network.

The second method whereby migration from subarea SNA to APPN may be accomplished is to use the composite node structure that was created in VTAM Version 4 and NCP Version 7. The composite node allows VTAM and NCP to create the image of a full APPN node. This node can support connectivity as either an EN or NN. This composite node is not an emulated APPN node, but a full node with a superset of APPN functionality.

This configuration allows the subarea SNA LUs to be registered to the APPN network and to support connected APPN LUs without restriction. If a session is initiated by the subarea SNA LU, the existence of the destination LU in the APPN network and the CDRM representing the APPN-side of the composite node must be predefined. However, the exact path or location of the LU is dynamically determined by the APPN network.

This is the most flexible configuration in connection options. It also supports the full APPN directory and dynamic topology services of any APPN node. There is little restriction on the path that a session can take. The TRS component of the composite APPN node can calculate a session path that takes advantage of the routes that exist within the APPN and subarea SNA networks.

A difficulty is raised, though, when attempting to migrate across network configurations, such as those of SNA network interconnect

---

[18]More than one node can be connected to the subarea SNA network, but because of limitations in NCP and VTAM, this configuration can result in complications.

(SNI). APPN is based on the paradigm of a global address space. Though all APPN nodes support connection to an EN across a network boundary, this mode of operation requires that the two connected nodes exist in networks that are adjacent or that share a common node within an adjacent network. This restriction could make it impossible for large subarea SNA networks to easily migrate to APPN.[19]

An APPN option set has been defined to allow a less restrictive configuration, but the architecture for this option has not been released. As a result, customers will have to rely upon IBM for all implementations that support a more flexible, architecturally correct option, at least until the option set is fully designed and is made available.

---

[19] The irony in this situation is that these SNA networks would also gain the most from the migration, because of the dynamic routing and discovery capabilities of APPN.

Chapter

7

Migration Methods for Subarea SNA to APPN

The previous chapter discussed the functional requirements of subarea SNA and APPN and what functions must be provided to allow a smooth functional integration of the two network architectures. That analysis included: where the two architectures provide common functions, what functions are unique to each, and how an environment of integrated actions can be provided. This could be thought of as a *what* needs to be migrated.

In contrast, this chapter will discuss operational questions of *how* the architectures can be integrated and/or migrated. This includes such issues as:

- The methods of providing a network that combines implementations of each architecture
- The steps necessary to provide an migratory path
- The impacts of different directions for migration

7.1 Requirements of Migration

The first step of any migration plan is to determine the requirements of the plan. Without a set of requirements for the migration, it is all but impossible to determine the correct choices for the migration, since you do not know what the parameters for the migration are or what the associated costs are for each of the choices.[1]

[1]I will proceed through an analysis of this migration. This may match with your requirements, but this is only an example. You must perform the same type of requirements definition steps to create a requirements definition for your installation.

Some of the requirements of a migration are:

- Compatibility of existing systems
- Support for all necessary functions[2]
- Support for a staged migration
- Concurrent operation of both systems

7.1.1 Compatibility of existing systems

Subarea SNA and APPN, though developed by the same company, have several areas of incompatibility. A migration from subarea SNA to APPN must deal with those areas and develop a method of overcoming them.

The incompatibilities of subarea SNA and APPN are centered around two areas. The first area is where APPN provides dynamic definition and the ability to find a resource in the network. The second area is where subarea SNA provides functions that exceed the functionality provided by APPN. For example, the lack of an MLTG in APPN is one of these areas where subarea SNA actually possesses a superset of APPN functionality.

A method of providing the maximum level of compatibility is one of the fundamental requirements of any network migration. In the case of subarea SNA to APPN, the compatibilities of the two network architectures is a strong requirement. This is because the network often represents a large investment by a company. In addition, the network may actually represent the lifeblood of the company.

7.1.2 Support for all necessary functions

The first step of this task is to define the necessary functions. This is not an easy task, since different areas of a company will often disagree on their view of "necessary functions." Often these differences are a reflection of each area's vested interests. An example of this is when an network management area designates network management as a necessary function, while an applications development group may not see this as their top priority.

One way to compile this list is to interview different groups within your company. By performing both a cross-reference and a tally of

[2]Support for all functions is desired, but support for those functions that you designate as necessary is a minimum requirement.

TABLE 7.1 Necessary Functions for Migration

Necessary Functions
▪ Support for sessions that cross the architectural boundary between the architectures
▪ No restriction on where sessions are initiated
▪ Support for sessions that pass from one architecture to the other and back
▪ Base-level network management
▪ Support for the dynamic functions of APPN, such as routing and network resource discovery

how often a specific function is mentioned, it is often possible to create a comprehensive list of "necessary functions."

Some of these functions will be common to all responses. These common functions represent the core of the necessary functions. Such items as illustrated in Table 7.1 are probably on this list.

Because both subarea SNA and APPN are session-based architectures, the session is of paramount importance to any migration scheme. The support for various session paths and types is required by any migration path between these architectures.

Assuming that you already have a subarea SNA network that is the target of a migration, support for all of the session types on your network is mandatory for the migrated system. You do not want to support certain interfaces that are only available under subarea SNA.

An extension of this is the requirement to impose no artificial limits on how those sessions can be initiated. Thus, it is necessary to support self initiations, initiations from the centralized datacenter, and third party session initiations.

It is also required that a session be started from either the subarea SNA or APPN networks. This allows sessions to be initiated by a resource in either network to a resource that is across the architectural boundary. This requirement allows free distribution of resources and allows all resources to be fully utilized.

The requirements should also include support for some of the extended APPN architectures, such as DLUr/s.[3] In addition, there should be no restriction on the placement of those dependent LUs. This requirement provides the flexible support for any dependent LUs within either network architecture.

---

[3]Although this is not part of the base APPN function, dependent LU support is so important because of the size of the potential resource pool. This is the reason I feel that this support is imperative.

7.1.3 Support for a staged migration

Implementation of a new network is seldom a simple endeavor. However, migration of an existing network is a complicated activity that requires a great deal of planning and support. If a migration required a complete swapping of resources and connectivity, very few migrations would ever occur. For this reason, the ability to support staged migrations is a requirement.

The migration of subarea SNA to APPN should be conducted in specific stages. The exact stages are dependent on the mode of migration, but in all cases, a staged migration is necessary.

It is possible to delineate the stages of a migration to ease the migration from subarea to APPN. These stages should be designed to migrate specific areas of the network. The network of these areas can be quite small in size to reduce migration difficulties. Two generic modes of migration are analyzed in Sec. 7.3.1.

7.1.4 Concurrent operation of both systems

Since a "light switch" migration is all but impossible except in the simplest of networks, it is imperative that the subarea SNA and APPN networks be able to coexist and operate concurrently. The concurrent operation of these networks is supported with few exceptions. As a result, you are able to migrate portions of your network while providing support for your users.

Some logical components of the network should be migrated in unison. If these resources are not migrated together, either the connectivity of the network will fail or the task of migration will become more difficult. These are resources that rely upon each other for operation. An example of this is the upgrade of VTAM and NCP. Although they do not have to be exactly coordinated, this migration should be done within a close time frame. In addition, if you are going to create a composite node, it is required that both VTAM and NCP be upgraded prior to proceeding.

In a similar vein, there are components that, if migrated together, would make migration difficult. These are resources that work in contrast with each other. An example of this is migrating CICS and migrating one of your applications to APPC. If you were to perform these migrations together it would make diagnosing a problem during the migration difficult.

Part of the design requirements for VTAM Version 4 Release 1 was that migration of the network be considered. The result was that VTAM supported most user configurations. The combinations of subarea SNA and APPN that were designed to be supported exceed the requirements of most users.[4]

One of the most difficult design points was to determine what part of the network to expose to the other side. During the design phase of VTAM and NCP, a decision was made to collapse the subarea SNA network and not expose anything, except for the network boundaries; the interior topology of the subarea SNA network is not exposed to the APPN network.

The APPN network can provide services to and utilize the resources of the subarea SNA network. With the advent of DLUr/s, the APPN network can even provide support for dependent LUs through a composite node. In addition, the composite node can provide the services of such functions as MLTG to the APPN network.[5]

Restrictions on the configurations that can be supported are a result of network designs that are not supported by the architecture. These restrictions are not specific to a migratory step. Instead, they are the result of restrictions in the architectures to support specific functions.

An example is the inability of an EN to provide intermediate routing. This restricts the configurations that a subarea SNA network can support in a migratory stage. However, this restriction is not the result of a restriction in the ability to migrate a configuration, but is the result of a design decision concerning the architecture. Because an EN does not include support for the ISR component, a subarea SNA network undergoing a migration is not able to provide this support. On the other hand, if the subarea components are in a different network topology, the migration is fully supported.

7.2 Methods of Migration

Migration from subarea SNA to APPN can be done in several ways. The method used is dependent on many factors that are unique to your requirements and configuration. The factors that must be weighed when making a decision on migration methodology include, but are not limited to:

---

[4]Of course, there is little doubt that among the thousands of subarea SNA networks that users support, many configurations were not accounted for, which may result in some restriction.

[5]An example of this can be seen in the VR-TG that exists in VTAM V4R2. The VR-TG allows the subarea network to fully utilize its MLTGs.

- Your network topology
- The time frame in which your migration is necessary
- Your available resource to support the migration
- Your budget for the migration

7.2.1 Network topology

The migration methodology is partially dependent on the topology of your network. If your network is small, the methods chosen may be different than if tens of thousands of devices must be supported. If your network is simple enough, it may be possible to perform a "light switch" migration that allows all nodes of your network to be migrated at one time.[6]

Another step along the progression of complexity is if your network contains a single domain. As processors have increased in capacity, more companies are reducing the number of processors that are required to operate. Thus, it would not be surprising that a sizeable network would be designed to operate in a single domain. Figure 7.1 shows a typical network that contains thousands of devices, but operates within a single domain.

In this case, there is no need to coordinate the migration of the host because the migration would be dependent on the functions that can be quickly migrated and the current release of NCP operating in the communication processors. But, even in this configuration, it is doubtful if migration can proceed as a "light switch" because of the large number of communication controllers whose NCP must be migrated.

If your network contains multiple domains, the NCP and VTAM upgrades must be coordinated. This makes the task of migration more difficult because more components must be migrated. In this case, the NCPs and VTAMs must be migrated and connectivity to the attached devices must still be maintained.

Another factor that will effect the migration path and speed is the use of independent LUs. The widespread use of independent LUs will make the migration easier. This is because the independent LUs do not require a definition of where they reside and how they will operate. These resources dynamically adjust to the current operation of

[6]Of course, if this were true, it is doubtful that this book will be of much use to you, as it would be surprising if you had a requirement for the integration of different network architectures.

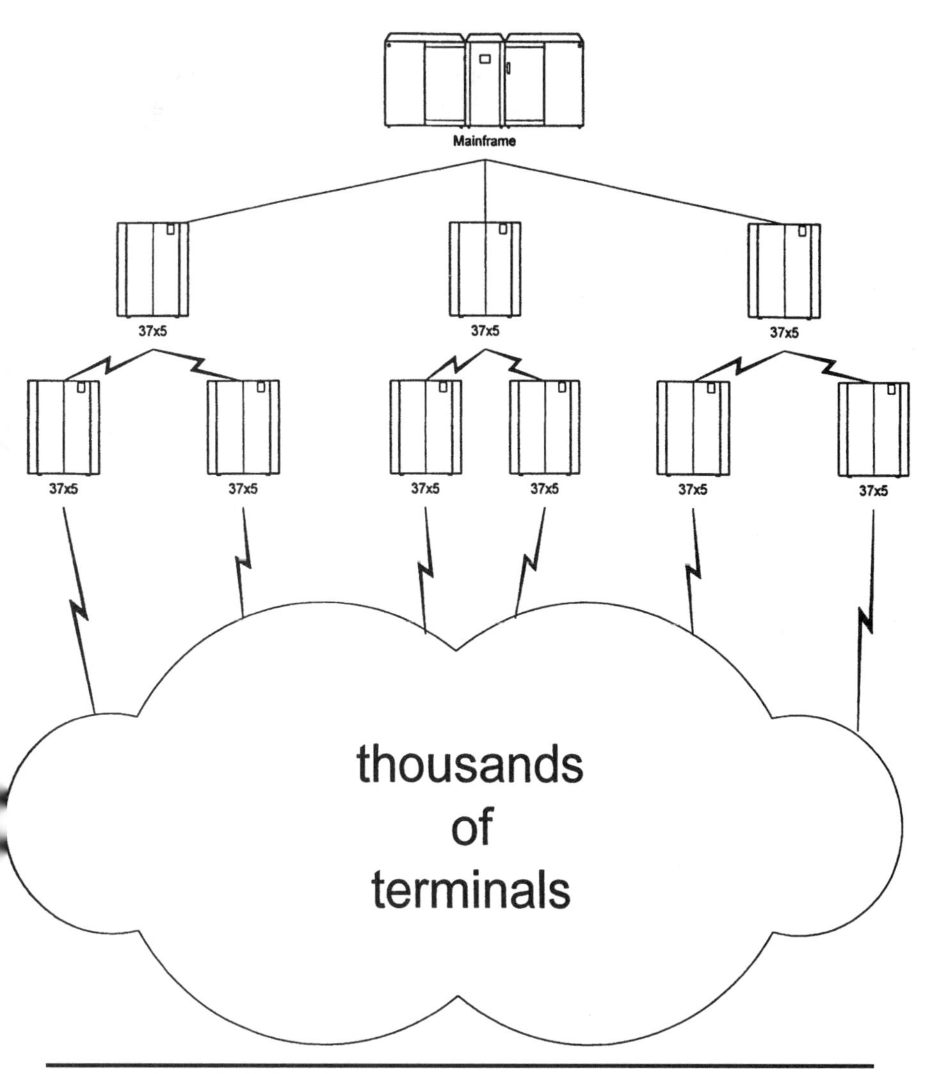

Figure 7.1 Large single-domain network.

the network. If APPN functionality is available, these independent LUs will utilize the higher level of functionality.[7]

[7]This method works as long as the resources are defined when supporting APPN. However, it is possible to restrict a resource to a lower level of functionality, but I am assuming that you have not done this.

Dependent LUs, which far outnumber independent LUs, also must be considered. If your network contains existing dependent LUs, you must consider how these are to either be supported or converted into independent LUs.

7.2.2 Time frame of migration

The way that you migrate is partially dependent on the migration time frames. This is based on the fact that a migration that is done quickly may require not only more resources to provide the technical support to perform the migration, but may also require additional costs.

These costs may be associated with contracting additional resources to support the migration. Assuming that your situation is similar to that of almost everyone else, you do not have many, if any, available technical resources. The ones you do have may be too burdened with supporting the existing network to provide either support or expertise for the migration to a new architecture.

Even through there may be costs associated with the increased speed of a migration, these may be offset by either software and/or equipment lease benefits, support costs, or just a matter of "having it over." These factors are very real and must be determined and resolved before starting the migration process.

7.2.3 Available resources to support the migration

Few companies have an excess of technical staff to support a migration. In almost all cases, companies have trimmed their staff in order to reduce costs and increase profit. If this is true of your company, you may find the tasks associated with performing a migration to be onerous.

Even if you do have people to support the migration, they may not have the necessary skills to provide a great deal of support for this type of migration.

The migration from subarea SNA to APPN is the movement to a new architecture—an architecture that few people fully understand. This type of migration can be quite difficult to perform, and it will be problem-laden. With the best of planning, the migration from subarea SNA to APPN is complex. Without sufficient expertise, the migration will be very difficult.

Knowledge of the architecture and the functions provided in the newest releases of VTAM and NCP must be fully understood before initiating this type of migration. When these steps are taken, it is possible to plan the steps to migration.[8]

7.2.4 Budget for the migration

As is true of many things we do, cost is a major factor in how your migration is done. Even with perfect planning and personnel, your available budget to perform the migration will effect your choice of migration methods.

The funding of a migration effort should not be overlooked as a key task of the migration. Without the proper funding, the migration will be difficult and will result in network outages and disruption.

One of the first tasks that must be funded is for staff training. Both the technical staff (system programmers, telecommunication analysts, and so on) and the operations staff (network control operators, problem desk staff, and so on) will require training on the new architecture and software features. They will also have to learn what options to specify in the software and the tools to support the network.

Only by providing this funding will your migration be successful. Planning is helpful in reducing costs by both speeding the task of migrating the network and eliminating many mistakes.

7.3 Migration

Migration of the basic network architecture is a complex task. As has been discussed, there are many requirements that need to be taken into consideration to plan for a migration. Now we will review some methods of performing this type of migration.

If the "light switch" migration is not possible, you are now left with the question, How should this migration be accomplished? The rest of this chapter reviews this question.

7.3.1 Methods of migration

There are as many thoughts about methods of network architecture migration as there are networks in existence. Each method has its own pluses and minuses. All methods must be tailored to the requirements of the specific network.

---

[8]Of course, you already know this, since you took the time to purchase this book to assist you in this task.

What can safely be said is that the migration must be a staged activity. The stages are the atomic migration steps that result in the entire network being migrated.

If only pieces of the entire network are migrated at a time, the important question becomes, what are these pieces? In the case of this type of migration, it is actually a complex answer that consists of many parts. Thus, the components of a migration consist of pieces at different levels.

One of the ways that a migration can be done is by splitting the network logically into either *vertical* or *horizontal* pieces. Another way to look at these options is to view the network as being made up of either vertical or horizontal components. Each of these components will be discussed in more detail.

The network migration, even if one of these high-level views is chosen, still requires key components to be migrated out of order. This is caused by many different requirements, including cost, key component, or the requirements of another project. When these "other" migrations are taken into account, you are left with a migration scheme that is often less than direct and also quite complex.

7.4 Vertical Migration

Vertical migration is based on the network view of vertical network sections that can be independently moved and migrated with minimal impact to adjacent and nonadjacent pieces. These sections are seen as being distinct from the others that are distributed throughout the network.

Figure 7.2 shows an example of a vertical migration. In this view, Network A should be migratable to APPN while Network B continues to use subarea SNA connectivity.

Between each of these nodes a *gateway interface* exists. This gateway provides a translation function so data can be transported between the networks.

This gateway function is very useful when migrating to an APPN topology. This eliminates the requirement to migrate all of your network to the new topology without any intermittent steps. Unfortunately, as we have seen, the migration path is not perfectly clean and clear because subnetworking is not fully supported.

One instance of this vertical network view is based on subdomaining of the subarea SNA network. This idea is based on the view of separate networks that are joined into a larger whole.

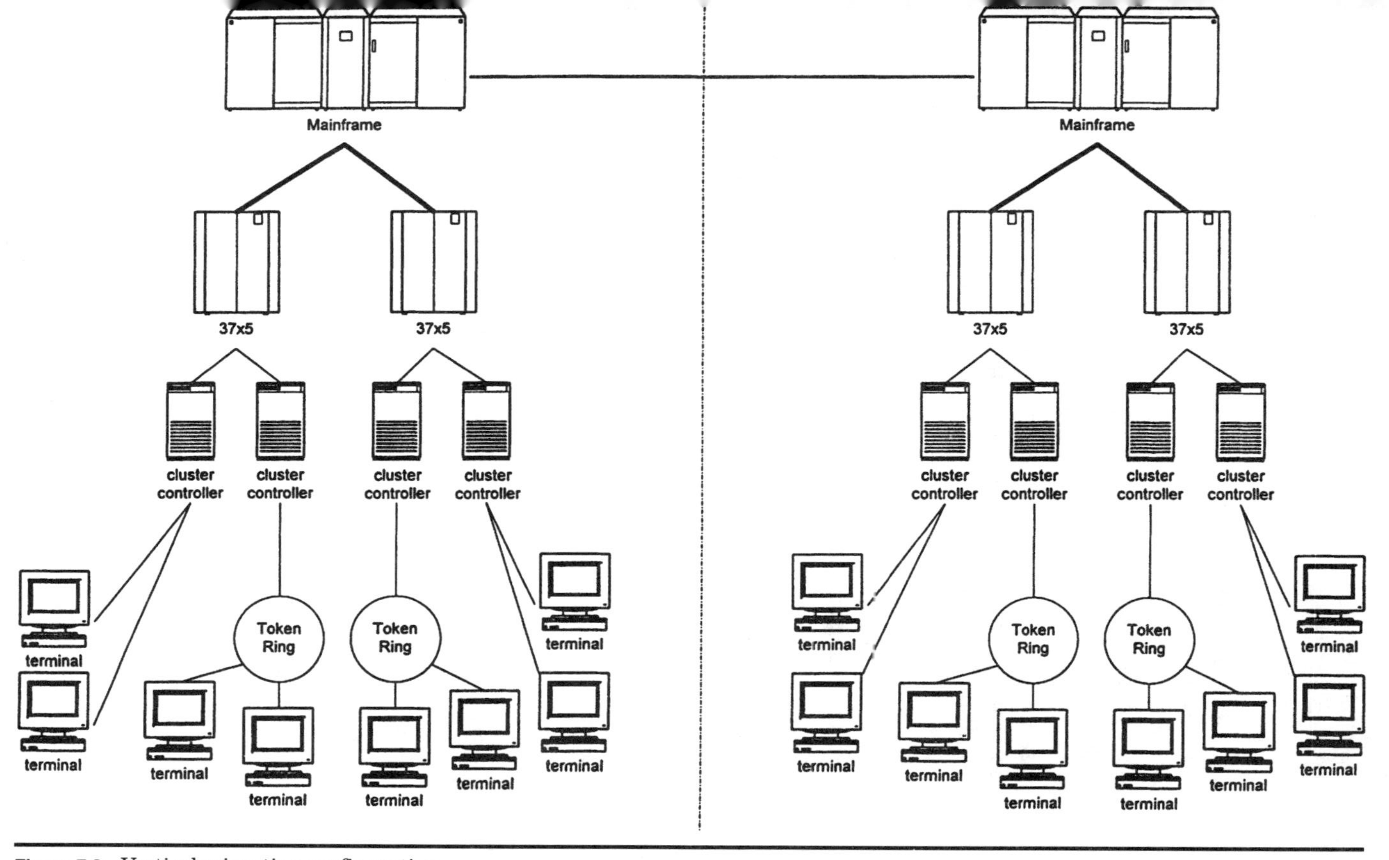

Figure 7.2 Vertical migration configuration.

7.4.1 Methodology

If you view the network as a set of stacked blocks adjacent to each other, you can see the basic view of a vertically segregated network. Figure 7.3 shows this view of the network.

Within this view of the network, it is possible to migrate vertical sections of network components. The basis of this migration method is to isolate the vertical sections from each other. Although the sections have a common connection at some point, they are largely isolated from the adjacent vertical section. If you do not change the horizontal interface between the sections, the adjacent sections will have no awareness that changes have occurred in the network.

At the same time, the sections that have been migrated can operate within themselves with all of the advantages of APPN. Full APPN directory and topology services are available within the vertical section.

For requests that cross the boundary to an unmigrated node, the migrated section emulates a resource type that can be understood. Figure 7.4 shows how two sections of the network can interface with each other. In this case, the two sections are quite different in implementation, yet allow communication between them.

One possible interface to provide this communication is emulation as a LEN node to the back-level section. By providing this type of interface, the back-level section can operate adjacent to the migrated section, while not imposing a large restriction on the operation of the network.

As was shown in previous chapters, this type of interface requires some system definition, but allows the two sections of the network to

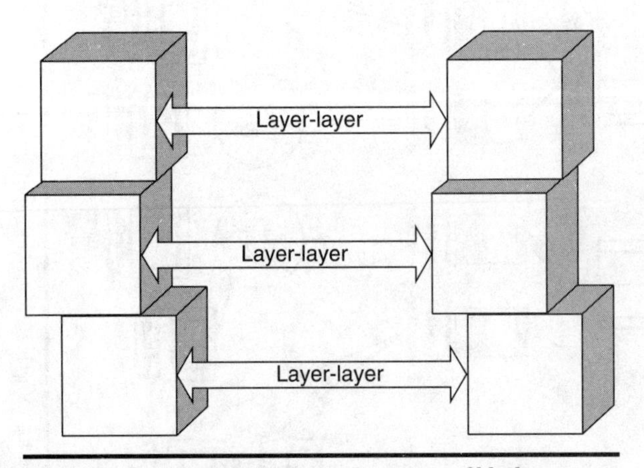

Figure 7.3 Vertical view of network as a set of blocks.

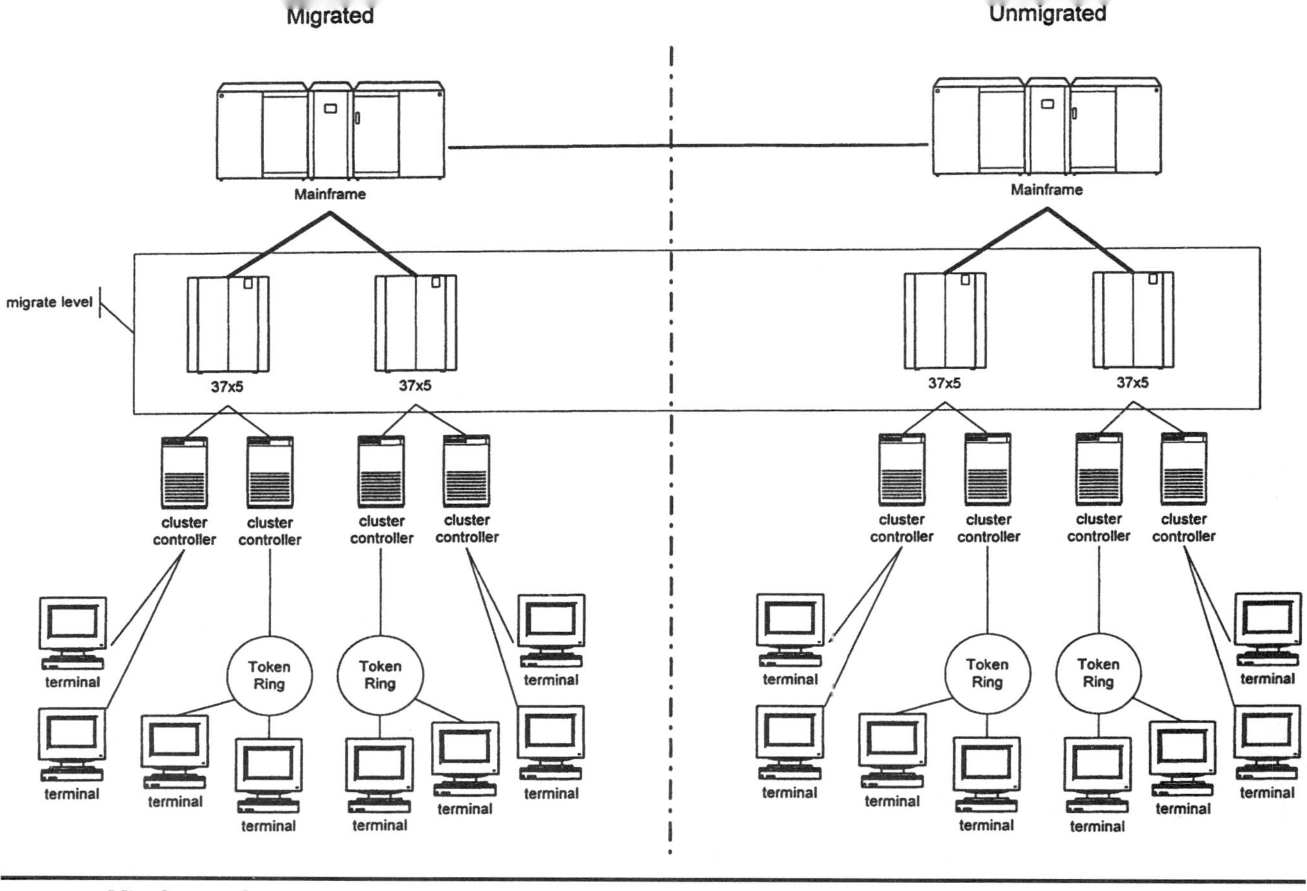

Figure 7.4 Mixed vertical migration.

communicate. Data can pass from either section of the network to the other. Although the network directory and topology services are more robust on the APPN side of the network, this should not have a major effect on the operation of the network.

Managing the network can be provided through the normal interfaces. As network management information passes freely between the two sections of the network, network management is fully supported. The only requirement is to define the management interface to the non-APPN section of the network. This allows the LEN-side to locate the network management application within the adjacent section of the network.

7.4.2 Advantages

If the pieces of the network are isolated from locally adjacent ones, then you can change one stack of blocks without modifying those blocks in the bordering stack. It is this stack isolation that makes this type of migration attractive. Using this method, you can make extensive changes to one stack without requiring changes to any other stack.

Assuming that requests for network services are generally localized,[9] most requests for network services will be in close proximity to the requester. Thus, there is a strong likelihood that a request for session services will be to the local host, i.e., the migrated host. An illustration of this locality of session is shown in Fig. 7.5. If this is true, the session service will use APPN rather than subarea SNA services.

Many companies use SNA network interconnection (SNI) in their existing subarea networks. SNI enables a company to cut the network into autonomous pieces that are separate from those that are adjacent. Each subnetwork is independent from its neighbor through the use of a *gateway node* that exists in both networks. Figure 7.6 shows how networks can be connected using gateway nodes at the boundary of a network. By using these gateway nodes, separate networks can gain connectivity.

In the case of APPN migration, the composite node can be seen as an extension on this view. Unfortunately, the composite node does not provide the same level of isolation as SNI.

The extended border node, implemented by VTAM V4R2, provides network isolation that is akin to that provided by SNI. This interface provides network isolation for any APPN node. Because only one node in the network needs to implement this APPN tower, by VTAM providing this interface, all APPN nodes within the network benefit.

---

[9]This locality paradigm assumes that most requests are for local resources.

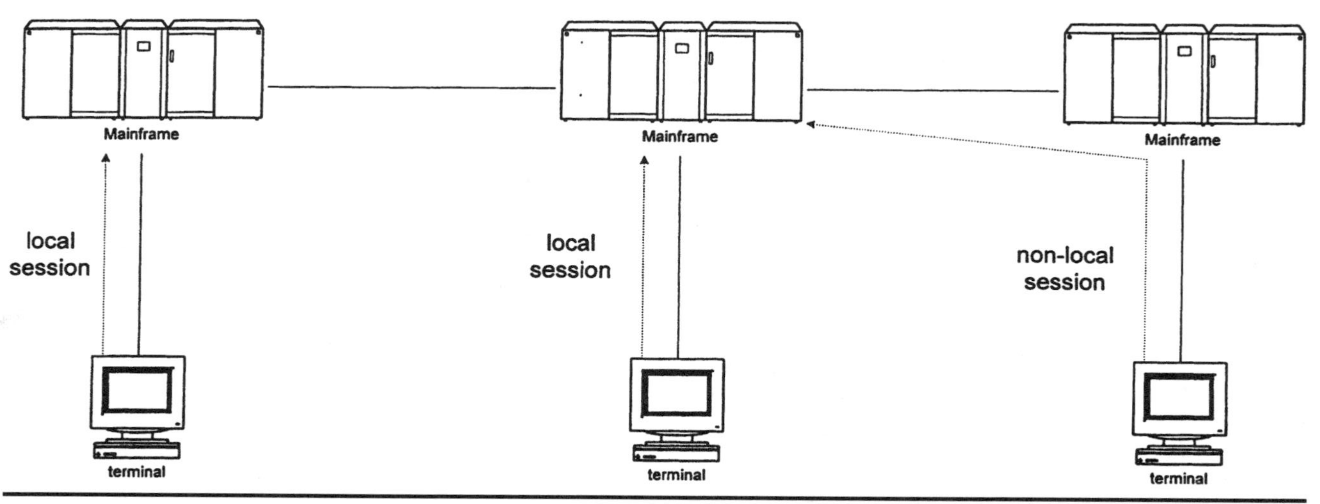

Figure 7.5 Local versus nonlocal sessions.

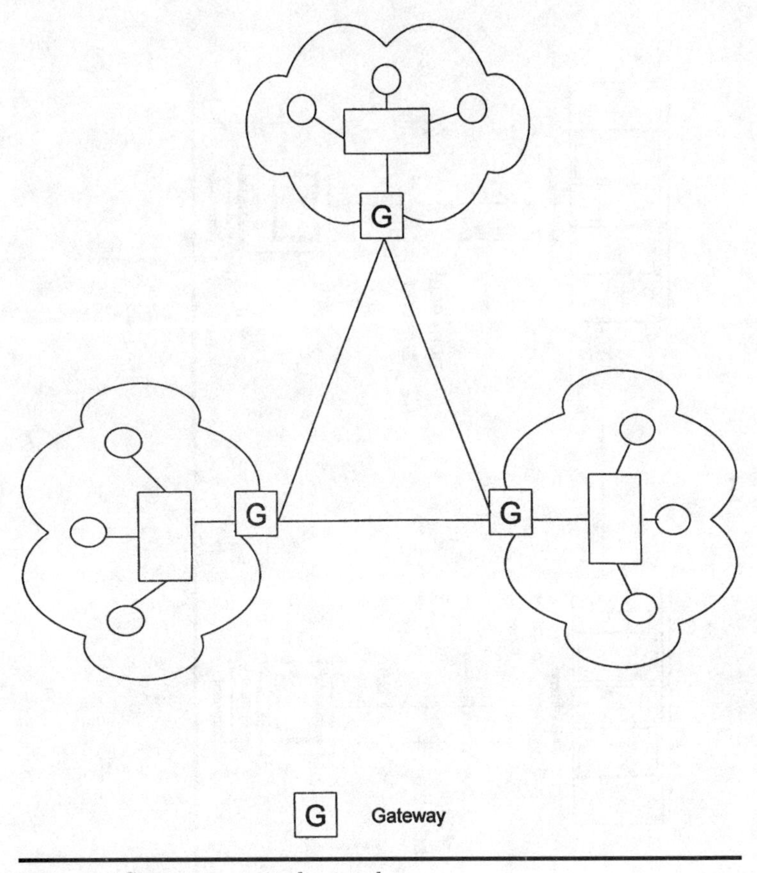

Figure 7.6 Gateway connected networks.

Because the architecture is not publicly available, vendors would have to reverse-engineer this support into their products.

7.4.3 Limitations

APPN is based on the idea of a flat address space throughout the network. Additionally, the boundary between network address spaces is very restricted in functionality. An example of this restriction is that a session cannot go further than one hop across a network boundary. This can be a severe restriction for many customers that are migrating from SNI environments, where the destination of a session can easily be more than one hop across a network boundary. The network boundary restriction provides support only when one of the session partners is adjacent to the network boundary.

VTAM has extended this restriction by implementing extended border node architecture in V4R2. If you do not have VTAM V4R2 in your network, you cannot utilize this attachment. In addition, the extended border node architecture has not been made publicly available, so no other vendor can implement the architecture.[10]

It is also likely that you will be required to provide a flexible migration path that steps outside of the strict vertical migration methodology. As was already explained, there will be instances where components within the network will have to be migrated out of order. When this occurs, then the strict vertical migration method will be violated.

This migration method also requires a mixture of network architectures within the "glass house." Thus, you will have a mixture of systems that will operate differently. You may have some hosts that operate as an APPN NN, another host that operates as an APPN EN, and others that still operate as subarea SNA network entities. In this case, operation of the network will be very complex because of the mixture of network architectures.

Some of the confusion results from differences in network management. Although both architectures can use NetView for management, the assumptions of network management between the two architectures is very different. The subarea SNA network assumes that all sessions are directed at resources in the host. APPN opens this up to other resources, such as peer level sessions. By having both types of network management interfaces, your Operations staff may become confused, which can have a negative effect on the perceived availability of your network.

7.5 Horizontal Migration

Subarea networks are often seen as consisting of logical layers. Figure 7.7 shows the same network topology as in Fig. 7.2, except that the view of the network is from a horizontal rather than vertical perspective. From this viewpoint, the network consists of logical levels that provide a certain service. Instead of vertical networks that are interconnected, the network consists of layers of concentration and functionality that are often interconnected only at the highest level.

---

[10]It is possible for a vendor to provide much of the architecture by using reverse engineering, but this does not allow a full implementation to be provided.

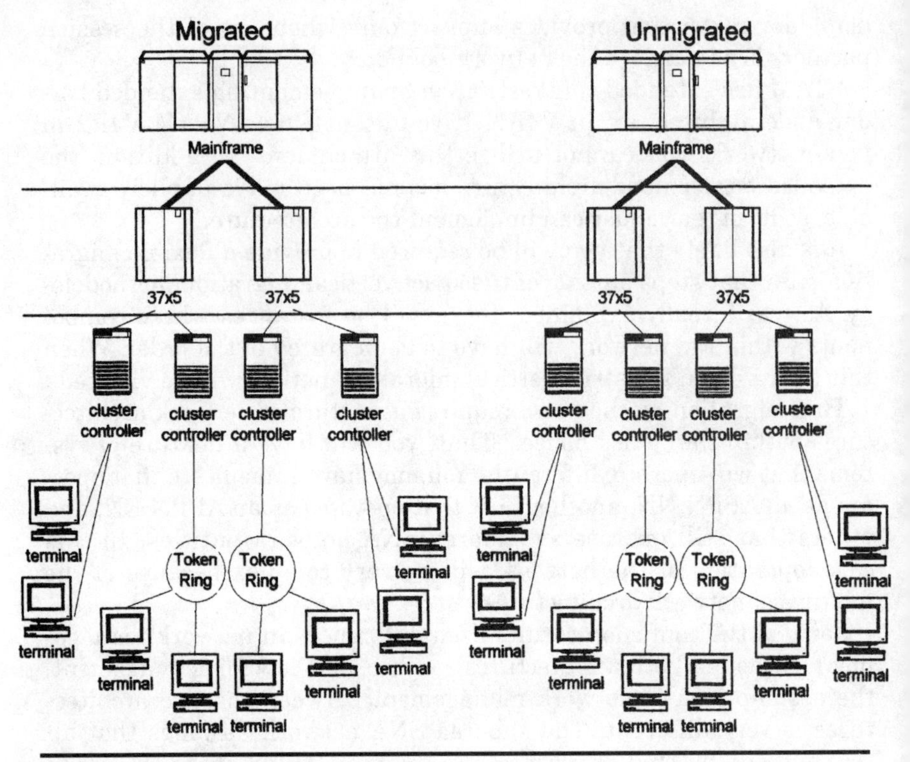

Figure 7.1 Large single-domain network.

This configuration can be thought of as analogous to the OSI networking model. Each layer has defined interfaces to the layers above and below it (this is like the vertical migration scheme). At the same time, each layer is communicating with a counterpart in another stack (this is the same as horizontal migration). These layers can be changed as long as the interfaces above and below are kept consistent. If the layer is architected correctly, this change is invisible to any other layer and to a partner layer.

When viewed in this fashion, it should be possible to migrate a layer of the network.[11] By migrating a layer, only specific interfaces are effected. This migration path is based on the idea that the changes can be better isolated within a layered network structure

[11]Often this technique is combined with vertical migration to further isolate the migration point, but we will not deal with this complexity in this book.

than in the larger vertical view. If the network is migrated from the top down, this view holds together well. Unfortunately, this requires that the largest and most heavily utilized portion of the network, the transport backbone, be migrated first. Though some companies will opt for this technique, obvious problems exist with this methodology.

7.5.1 Methodology

A horizontal migration is provided by picking a logical level of the network and initiating migration for that level. Though migration can be done at multiple levels concurrently, the purpose of this type of migration is to complete a major portion of a level before commencing migration at a new level.

In order to succeed with this type of migration, you must plan carefully. This planning must provide for the following:

- Which level to migrate first
- What steps are required before migration is feasible
- What impediments to success exist
- Time frames for migration

Once these steps have been taken, you must also provide the exact steps necessary for the migration. These plans should include provisions for any restrictions to the migration and key components that need to be migrated to lay the groundwork for other stages of the migration.

A *complex migration* scheme[12] can be reviewed during the planning for a migration. A complex migration entails migration at multiple levels concurrently. This can be viewed as migrating at both vertical and horizontal directions in parallel. The coordination necessary to provide an orderly migration is also complex.

7.5.2 Advantages

Horizontal migration has the advantage of requiring less widespread migration than vertical migration. This is because the layers of a complete stack do not need to be migrated together, or at least in close proximity. The layers can be migrated in the order that makes sense logistically, politically, and technically. There is no requirement to coor-

---

[12]A complex migration scheme is a feasible extension to this methodology, but goes beyond the scope of this book.

dinate the migration on a vertical basis, which may require the intervention and involvement of a larger percentage of the technical staff.

A horizontal migration allows for isolated sections of the network to be migrated without the direct and immediate involvement of other sections of the technical staff of your company. Assuming that the layers are isolated, this type of migration can be done with a minimal amount of technical staff involvement. Using horizontal migration, even a single location or component may be migrated. Although the immediate benefit of this migration will not be realized, this migration flexibility allows for easier coordination and planning of a migration path.

As stated in Sec. 7.4.3, even in the case of vertical migrations, in almost all cases there are requirements providing for the horizontal migration of some components. The opposite is not necessarily true; there is no requirement to vertically migrate a section of the network, because it is possible to migrate an installed network without any requirement for vertical migration. Thus, by proceeding on a horizontal migration direction, it is possible that a complex migration scheme may not need to be created.[13]

7.5.3 Limitations

The horizontal migration method does not result directly in a section of the network being migrated. Because each piece of the network is viewed independently, it is possible to create a hodgepodge of isolated migrated islands that results in little or no advantage for the users, operators of the network, or the associated costs of providing and maintaining the network.

As a result, this migration method, like the vertical method, requires strong requirements and planning steps to ensure that the migration is moving forward and that a staged migration results in advantages. Without realizing its advantages, the support, operations, and administrative staff will view the migration as a costly chore that does not result in any benefits.

There are also times and configurations requiring a strong coordination of migration steps on a vertical level. An example is a configuration that contains a large portion of dependent LUs that will require support of DLUr/s. Since this support was not available until VTAM V4R2, vertical migration steps are required.[14]

---

[13] I qualify this statement with a large "possible"! In almost all migrations I have been involved with, a complex migration scheme has been necessary. As a result, this is more of a theoretical statement, rather than an expectation.

[14] If this is not done, it is possible for these dependent LUs to not have the required support if VTAM is not migrated to the proper level.

7.6 Network Management

Network management in this environment can be a challenge. If your network operations staff is already accustomed to subarea SNA management, the transition will be somewhat difficult. Although the same applications may be used,[15] the assumptions of each architecture make the migration of the staff troublesome.

For example, assuming that your staff is accustomed to subarea SNA, they will assume that the host has awareness of all devices in the network. Because this is the case in subarea SNA, your staff may continue to view the network in this manner. But reality shows that in APPN, VTAM has no awareness of a specific destination. This is because the location of that destination is not predefined. Once that destination contacts VTAM, its location will be cached and known, but until that time, the resource is in an undefined state.[16] Thus, the network staff requires some training to gain an understanding of APPN and its implications to their work.

When using DLUr/s, the downline PU and LU are actually disconnected from the upline host. In this configuration, a network management application, such as NetView, can provide a misleading picture of the network as being contiguous. Work is required in this area to provide further refinements to the architecture to allow these complex management environments to be effectively pictured and managed.

One way of extending management is to provide a hierarchical mapping through the use of the FQPCID. Since this field is designed for indexing into a management table, it would seem appropriate to use this field as the index method for network management. The network name, NETID, and other "normal" human interfaces can and should be used to enable maximal management support.

Another aspect of this problem can be seen when connectivity to the remote PU/LU is lost. The network management application may provide good information on the cause of the problem, but it is also possible to obtain little or no information.

The DLUr could alternatively take on the role of a *service point.* This might be used to provide a standardized method of interfacing into the session between the DLUs and DLUr. This activity could be done so that the network management application can obtain management information within the LU 6.2 session. This is a fully programmatic interface that can be driven by automated tools, such as CLISTs.

---

[15]For example, NetView is compatible to both subarea SNA and APPN network architectures.

[16]VTAM attempts to locate the report on the resource. However, this method takes longer and has a greater effect on the network as broadcast LOCATE requests are sent out and responses arrive.

7.7 Conclusion

Migration of the network from subarea SNA to APPN is not an easy endeavor. It requires a great deal of planning and testing of interfaces to ensure your users as little interruption of service as possible. The steps of the migration should be small enough so that each migration stage can be done quickly and a fallback plan should exist for each migration stage.

As part of the planning stage, you must decide on the fundamental migration scheme that will be used. I have sketched out two very high-level methods, but you may find that your requirements demand a different scheme.

The vertical migration method is based on the assumption that a network can be divided into vertical segments that can be acted on independently. It also assumes that the migrated segment of the network can gain advantages by being migrated.[17]

The horizontal migration path provides a migration of a layer of the network. By migrating a layer, it is assumed that this layer can be isolated from the layers above and/or below. As is true of the vertical migration scheme, it is possible that it will be difficult to provide the necessary isolation, dependent on the specific configuration that needs to be supported.

Other necessary migration plans are needed to support some subpieces of the network. An example is the migration of the dependent LUs of subarea SNA. Since APPN provides support for only independent, LU 6.2 sessions, the millions of dependent LUs are left out. As a result, a new architecture, called DLUr/s, was created to lend them support. This architecture allows the data flows of dependent LUs to pass through the APPN network. In this configuration, the APPN network becomes the medium of transport for the dependent LU flows.

---

[17]This is not an assumption that can be easily made, since advances from the migration are subjective, rather than objectively demonstrable.

Chapter

8

Methods of Integrating SNA and IP

This chapter provides the first glimpse at the task of integrating the SNA architectures, subarea SNA, APPN, and HPR and the IP network architecture. As reviewed in Chap. 5, the SNA and IP architectures are based on different assumptions about the network and the responsibilities of each component within that network.

These differences and assumptions make the task of integration a challenge. The complex part of this task is the profound differences between the architectures. The divergence of the architectures must be overcome by some means. As the architectures have sharp contrasts in their operation, addressing, and recovery procedures, either a common interface must be provided that masks the differences or a means of providing an emulation for one architecture to provide the abilities of the other must be provided.

A method of finding a common ground for these architectures is a complex task that requires an understanding of each network architecture to be integrated, and devising a means of creating a common interface that provides a groundwork upon which each architecture can exist and operate.

The available methods for providing this common base are to use a hardware or software interface that facilitates the creation of a common groundwork for each architecture. This groundwork can provide different types of interfaces. The differences between the interfaces are based on several levels of differentiation. These include:

- Cost of implementation
- Features provided by the interface
- Complexity to establish the interface

It is important to remember that there are reasons to use one of the interfaces; the attractiveness of one interface does not eliminate the use of any other interface. You must define your requirements, budget, time for completion, and experience of staff prior to determining which interface is best for you.

Often, a protocol is dictated by the design of an application. For example, if an application is built to use a sockets interface it will almost always use an IP-based network. Though there are several alternatives to the use of IP, it is unlikely that they will be used.

In a similar vein, if an application operates in a CICS, it is likely that it will use SNA protocols. This is in spite of the fact that CICS operates in environments, such as an AIX-based RS/6000 or on an OS/2-based PC, that are not normally SNA environments.

Thus, the choice of protocol is not only a matter of what protocol is desired. Often the environment of the desired application leads to the choice of a particular protocol. With this in mind, it is required that diverse protocols have an environment in which they can coexist.

8.1 Hardware Techniques

Hardware techniques have the ability to use some hardware interface to provide the necessary groundwork for the integration of these architectures. These hardware techniques range from the use of additional hardware interfaces to the use of a multiplexing technique.

The one side of this, additional hardware interfaces, can be thought of as analogous to adding additional lines to provide connectivity to an additional endpoint. Though this may not be the most economical, it is a perfectly valid method of providing that connectivity. This would be especially appropriate if the additional architecture is isolated from the others and if your budget has sufficient funds to provide support for this technique.

8.1.1 Multiple physical paths

Figure 8.1 shows an example of the use of multiple hardware interfaces. In this example, the TCP/IP and SNA resources are linked through the use of multiple physical interfaces. Though this requires the use of multiple interfaces, and is thus not the most efficient method, it is a valid example that should not be short-changed. The use of this type of integration could be used for many reasons. Among these are:

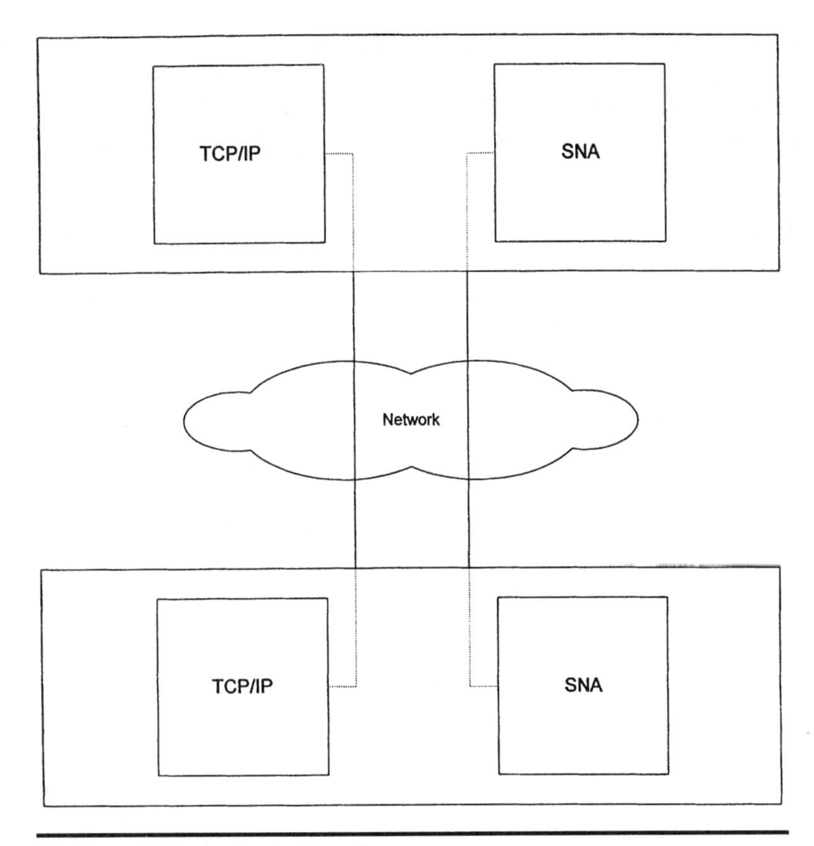

Figure 8.1 Multiple physical interfaces.

- *For short-lived interfaces.* It is possible that one of the interfaces is only required for a short period of time and will then be migrated to a more consistent interface. In this case, the use of a migration stage is built that will lead towards its own obsolescence.
- *An interface that has other reasons to be isolated.* There could be other valid reasons to have an isolated interface, such as security requirements.

If a resource is in the process of being migrated, there is little reason to provide the most efficient interface possible. In this case, resources are better directed toward supporting a more stable interface. If an additional interface is maintained for a period of time, this may be the most efficient use of one's resources. The cost of creating a single interface through which the SNA and TCP/IP interfaces could operate may exceed the cost associated with having multiple physical interfaces. In this case, the use of multiple interfaces may actually be less than the cost associated with the cost of creating a merged interface.

There also may be extenuating circumstances that make the use of an isolated interface attractive. One example of this is, if there is a security requirement that makes the use of an isolated interface a benefit rather than a liability. In this case, the use of an isolated interface is an attractive alternative.

8.1.2 Multiplexing

Another alternative hardware solution is to multiplex multiple lines across a single physical interface. This solution results in the use of only a single line to carry the data traffic from multiple origins and supports more than a single protocol. Fig. 8.2 shows an example of the use of hardware multiplexing.

This solution can be quite useful, if the separate protocols end at a single destination. This is a simple solution that results in a savings of additional data lines. Although it does require the use of two multiplexers, this is not usually an exceptionally high cost that would eliminate this as an economically viable solution.

The use of hardware multiplexers requires that two hardware boxes be used; one for each end of the circuit. Each end takes in the multiple protocols and multiplexes the data traffic across a protocol that supports the transmission of the underlying protocols. This protocol is an HDLC derivative that is often proprietary. At the remote end, each data path is separated out and transmitted to different physical ports.

Multiplexers come in several operational models. The first is a *frequency* multiplexer. This type uses different frequencies to multiplex the different lines. Each data path uses a different set of frequencies to allow the multiple data streams to be supported simultaneously by separating the traffic across different data paths.

Another type of multiplexer is a *time division* multiplexer. Each data path is multiplexed by allocating each datastream a fixed period of time to transmit. The multiplexer divides up the bandwidth into fixed blocks of time. Each datastream is transmitted for only that small period of time before the next datastream is sent. The streams are, in essence, packetized into fixed-length blocks with a header identifying which datastream is contained.

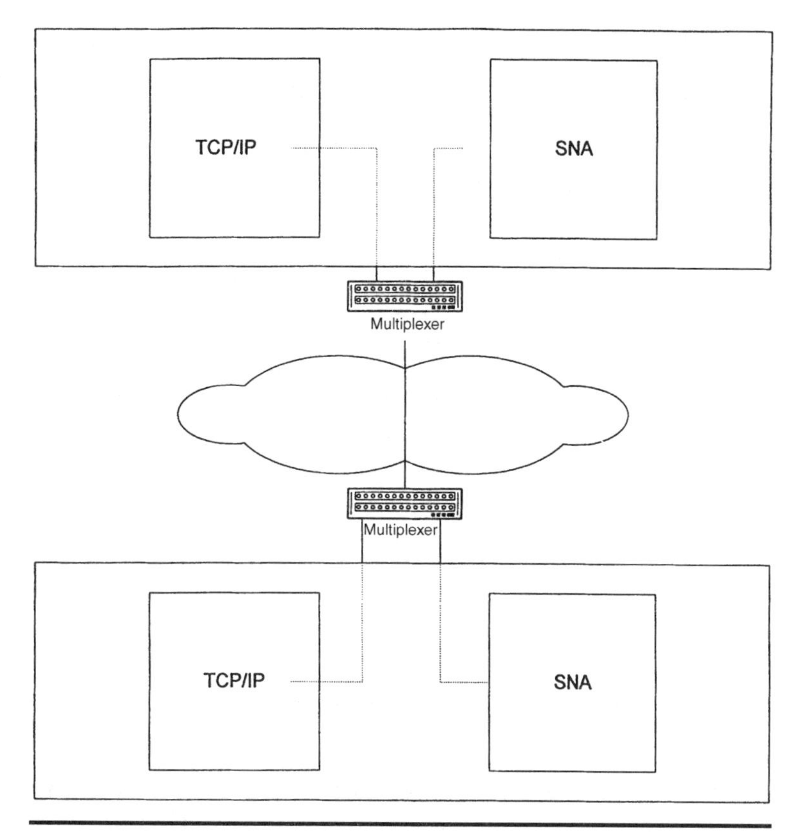

Figure 8.2 Multiplexed protocols across a single interface.

Each data path shares the physical circuit in complete fairness. Figure 8.3 illustrates the use of this type of multiplexer. You can see in this figure that each data path is given the same amount of time to transmit data.

At the remote end, the data paths are separated into the original data streams by using the inverse of the multiplexing algorithm. The reconsolidated data paths are output to separate data ports connected to different destinations.

The problem with this type of multiplexer is that it is too fair. Normal communication analysis shows that "some origins are more equal that others." This means that the origins and destinations are not equally utilized; some have more traffic than others. The use of a time division multiplexer does not take these differentials into account.

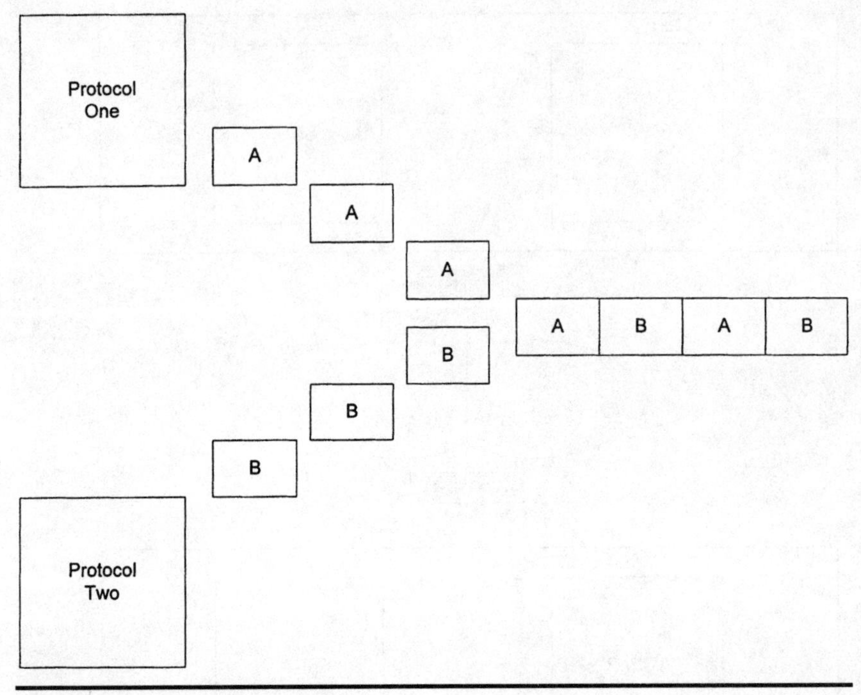

Figure 8.3 Time division multiplexing.

The last type of multiplexer overcomes the shortcoming of a time division multiplexer. This is the *statistical multiplexer.* The stat mux, for short, analyzes the data paths and determines the use of each data path's time and adjusts the time slots accordingly. Thus, a busier port is provided more time slots than one that has little traffic.

This adjustment is done on a continual basis, dependent on the amount of traffic at a single point. This allows adjustments for the changes in the use profile of different ports. The adjustment provides for a dynamically adjusting network that keeps the physical circuit as busy as possible.

8.2 Software Techniques

There are also software techniques for the integration of multiple network architectures. The techniques available can be grouped into four methods:

- Multiplexing of protocols on a common base
- The encapsulation of nonsimilar protocols by a single protocol

- The emulation of nonsimilar protocols into a single protocol base
- The conversion of nonsimilar protocols into a single protocol base

8.2.1 Software multiplexing techniques

The software multiplexing methods are similar to those used in hardware. These methods, like the ones used in hardware multiplexers, provide a common base upon which multiple protocols are carried.

The simplest method used for software multiplexers is the same as in hardware multiplexers. These products are software versions of the a hardware emulator. There is a tradeoff of using a software rather than hardware multiplexer; these products are less expensive, but are also slower in operation. A software multiplexer is also more flexible for upgrades and modifications.

These products provide multiple queues for inbound and outbound data. The line scheduler scans the outbound queues for data to be transmitted. A small header is added to the data to identify the input and output queues at the respective ends of the line. As is true of hardware multiplexers, these products use different algorithms to determine the order of transmission and differentiate them from each other.

At the other end of the spectrum are the current advanced "multiplexers." These products use various architectures to provide a multiplexing solution that can interoperate with other products, both software and hardware based.

8.2.2 Encapsulation of nonsimilar protocols

As described earlier, one of the methods of supporting multiple protocols across a single connection is to use an *encapsulation* technique. Encapsulation provides for the creation of an envelope into which other protocols are placed. The data within the envelope is not seen by the network and has no effect on the network. This facilitates the transmission and reception of data in a form that would normally not be recognized nor supported.

Once the data has arrived at its destination, it is removed from the envelope. It then assumes its original form or serves its original intention.

This is analogous to mail delivery. Mail is picked up and delivered without knowledge of the contents of the sealed envelopes that are transported. There is no awareness of whether the contents are a note to a mother, a love note, a threat, or a check for one million dollars. In all these cases, the mail is transported and the anonymity of the contents is kept.

Examples of encapsulation methods include several IETF RFCs. These include RFC 1490 and RFC 1356. The purpose of these methods is to provide a flexible, consistent method of supporting multiple protocols that can interoperate with other implementations of the same RFC.

RFC 1490 uses a frame relay network as the common base upon which the other protocols are passed. This RFC uses a variable-length header that identifies the type of data that follows the header. As Fig. 8.4 shows, this technique provides exactly the same type of datastream as is seen coming from a hardware multiplexer. The data is passed as small packets that are identified through the use of a header.

The receiving end takes the packets off of the line and reassembles the pieces into the original messages. This reassembly is based on fields within the header. The header identifies the origin, destination, and some characteristics of the data. The receiving station uses this information to assemble the message into its original form. It also uses this information to complete the routing of the message to the

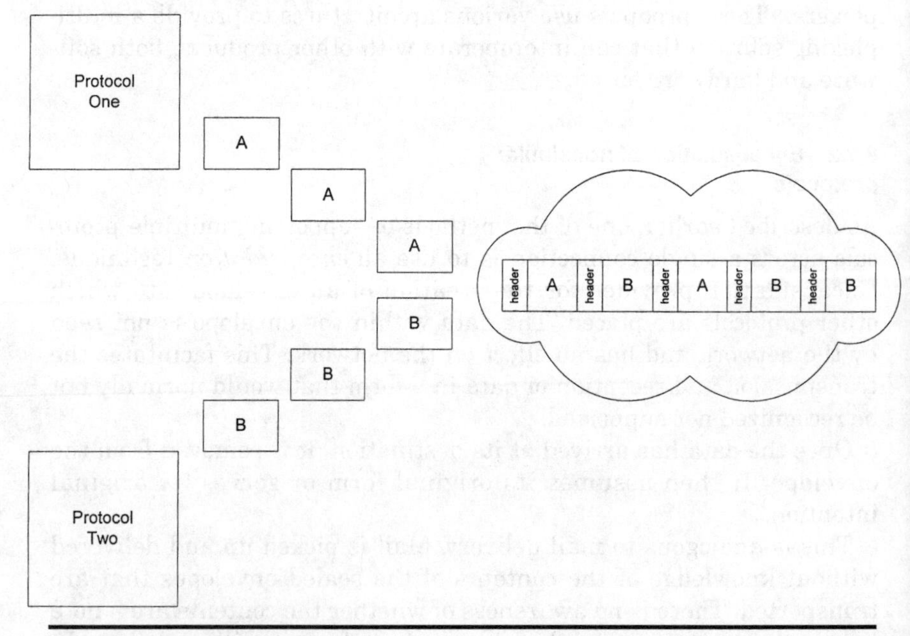

Figure 8.4 RFC 1490 multiplexing.

correct destination. This process is very similar to that done by hardware multiplexers.

8.2.3 Emulation of nonsimilar protocols

Another method of providing support for multiple protocols is to terminate one protocol at the boundary of the network and pass the data on a different common protocol. This emulation requires two emulations. The first is the termination of the noncommon protocol, while the second emulation is of the common protocol. This allows the user data to be transported while not requiring support for a new protocol. This emulation provides an interface for the data within the foreign protocol to enter, and be transported by, the network. Figure 8.5 illustrates the method by which emulation allows the transportation of data from a noncommon protocol.

Many types of emulation can be utilized. The type that is best for your purposes is largely determined by the protocols that you want on your network. There is no "right or wrong" protocol or emulation. At the same time, some emulations are more prone to problems than others. This is because some protocols are more sensitive to certain types of changes in the network. For example, the SNA protocols are more sensitive to timeouts than the IP-based protocols. As a result of this, SNA emulations should be used with care.[1]

[1]Many SNA emulations provide the ability to successfully terminate non-SNA protocols. This direction can be taken, but proper care and testing is paramount to a successful implementation.

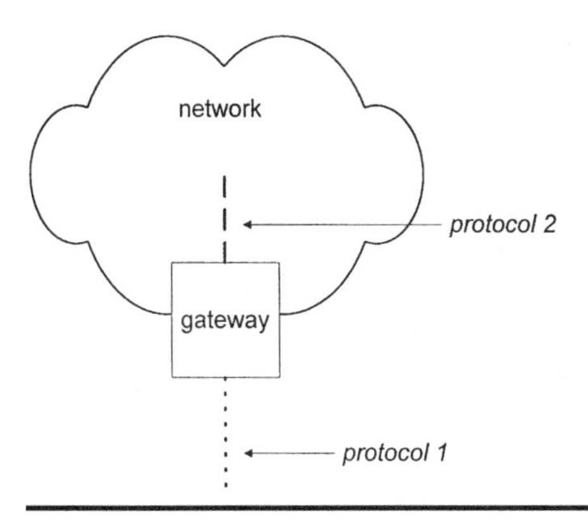

Figure 8.5 Conversion of protocol.

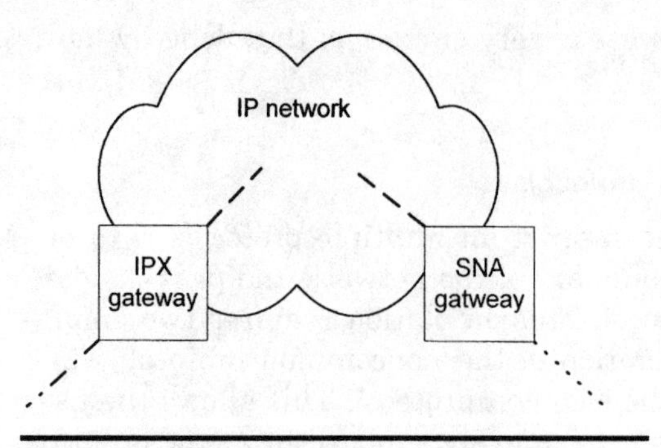

Figure 8.6 Conversion of noncommon protocol to single common protocol.

It is also possible to emulate multiple protocols so that the result is a single common one. For example, it is feasible, even if there are only two protocols entering the network, to provide emulation for both of them. In this configuration, the desire is to have a network that uses a different protocol than either of the ones that are entering the network.

An example of this is shown in Fig. 8.6. In this case, IPX and SNA are entering the network, but the foundation of the network is IP. Both of the incoming protocols are emulated into an IP implementation upon entering the network. The network only supports IP, in this case, to ease both support and management complexity. At the other end of the network, the emulated protocols are returned to their original form, for delivery to their final destination.

8.2.3.1 Emulation of IP by SNA environment. Several SNA-based environments provide for an IP emulation. These products range from implementations of IP on personal computers to IP operating on the largest mainframes. In many of these cases, the IP transport can provide a gateway function between the IP and some other protocol. The purpose of this interface is to provide a method of passing data from one environment to the other.

The level of integration between the two environments can be very different for different implementations. In some of the implementations, the integration is not mature. These implementations exhibit a lack of simple transference between the protocols. The transmission of traffic is provided in a store-and-forward manner.[2]

[2]There are few implementations that provide this almost total lack of support in existence at this time.

Other environments provide a full interchange for a variety of the IP-based protocols. These environments usually include TCP and UDP data traffic. This traffic is passed onto SNA LU sessions for continuation to the destination.

One of the greatest difficulties of these gateway products is address resolution between the two architectures. The method of addressing an origin and destination of a message is very different between subarea SNA, APPN, HPR, and IP. As such, it can be a true challenge to provide address resolution and conversion between the SNA architectures and IP.[3] Usually an address conversion table is created. The basic form of this table is two columns. The first column defines the IP addresses (or IP address range) and their associated SNA address.[4] This simple table allows the software to make the translation between the two address methods.

8.2.3.2 Emulation of SNA by IP environment. Just as SNA environments have the capability to emulate IP, IP environments often have SNA emulation available. These emulations allow IP traffic to be passed onto SNA sessions and passed to an SNA host.

In most of these cases, this emulation provides the image of a cluster controller and associated terminals to the SNA host. This type of "terminal emulation" does allow for connectivity, but does not provide the best connectivity options. The throughput of this type of interface will not be the most optimal. In addition, the connectivity options are limited. This is because of the limitations of the "cluster controller" type of interface. The limits include a single access point into the SNA-emulated host and limitation that data is only transmitted upon being "polled from the host."

Another issue with this type of interface is that the subarea SNA emulation can be error prone and may fail from an inability to fully emulate the complex SNA protocols. When this occurs, you will typically have the choice of pointing the finger either at the emulation software vendor or at the SNA host vendor. Since these situations are less than entertaining, it is not unusual for customers to attempt to stay away from such implementations.

Another alternative SNA emulation is the newer APPN node types. Though several vendors have announced development of these interfaces, the implementations are still fairly immature. Since this archi-

---

[3]See Chapter 5 for more discussion of this topic.

[4]An SNA address in this type of definition is a PU/LU address pair. This provides the desired dynamic definition and allows for a fairly simple setup.

tecture is still being developed in many areas, the implementations have had to lag behind. As such, some of the emulations need complete testing and tuning for throughput.

At the same time, the existence of vendors providing APPN implementations is itself a marked step forward. This is because of the greater flexibility and capabilities inherent in the architecture. By supporting APPN, vendors are able to participate in directory and topology services provided by the APPN architecture. Thus, resources are able to be dynamically located and the network topology is dynamically built, instead of needing predefinition as in subarea SNA. This can greatly aid the customer in providing dynamic routing—inherent in IP networks—within the SNA world. If the translation between the architectures is done well, it is possible for resources to be located that cross the translation boundary. This smooth translation can provide an easy method for customers to gain communication from IP resources to APPN and subarea SNA resources. This eases this type of integration so that the movement between these architectures becomes almost seamless.

8.2.4 Conversion of nonsimilar protocols. As I previously discussed, it is possible to convert different protocols to other protocols. Although this would probably not be the normal direction to take, it is a perfectly acceptable option and one that can be useful in some circumstances.

For example, suppose the protocols arriving from the network are 3780 bisynchronous and polled asynchronous. Although the network is delivering these protocols to the boundary of your network, you may not want to add support for these protocols[5] to the interior of your network. As a result, you need a protocol converter that will convert each of these to a more generally supported protocol, such as IP.[6] The protocols that you do not want to support within your network are converted into something else at the boundary and transported through your network in the converted form. In this case, these protocols are not seen within your network. Figure 8.7 illustrates this conversion.

---

[5]For that matter, you may not be able to find interfaces that support these protocols.

[6]There are several companies that can supply these types of "gateway" products. Among the companies that support them are Hypercom and Novell.

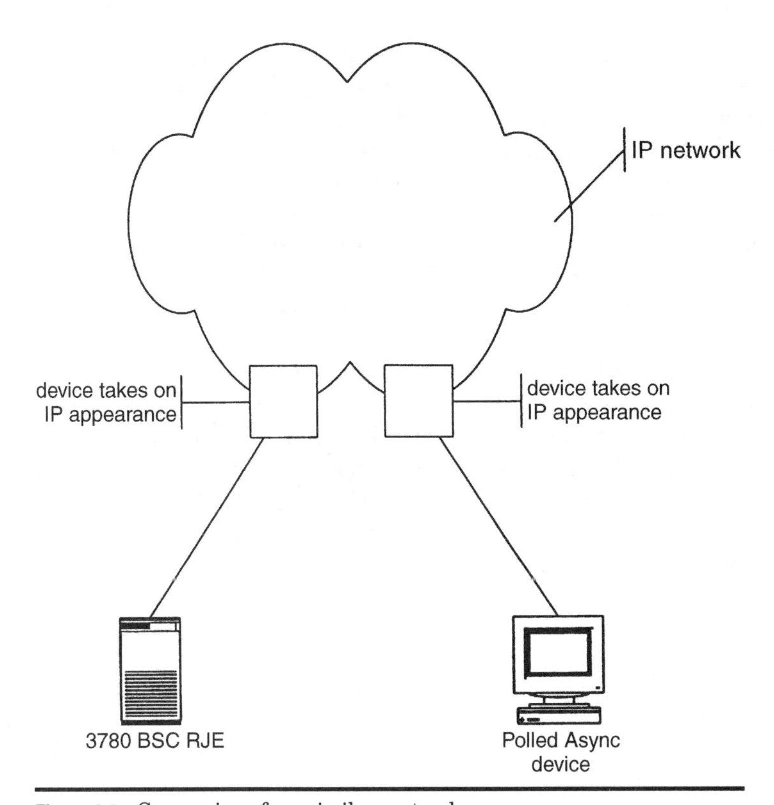

Figure 8.7 Conversion of nonsimilar protocol.

8.3 Standards-Based Integration

Because of the great demand for either the integration of different protocols or connectivity to nonnative protocols, several standards[7] have been developed. These standards allow the support of protocols that are not native to the environment to which they will be attached.

The support for connectivity to nonnative networks results in data being able to pass into your network. The other interface is seen as being external to your network, but data is able to pass between the points of the network. An example of this is if an IBM host wanted to allow connectivity to an IP-based network, it could use one of these standards to provide that support.

[7]The term *standard* is used a little loosely in this context. What I mean is that there is an accepted and widely supported method of providing that support. This equates into a de facto rather than voted-upon standard.

It is also possible that what is desired is a true integration of the nonnative protocol into the existing network. This would require a higher level of support than required to simply allow connectivity. In that case, the outcome is an integrated network that does not recognize the existence of a different interface. Instead, with true integration of the nonnative protocol, you see a contiguous network that provides connection and integration for several boundary connections to your network. The nonnative interfaces are not external to the base network, but rather an *extension* of the network to areas that it did not reach previously.

The point of commonality of all of these standards is that they are open standards that either are, or could be, implemented within a diverse set of environments. As such, they are not proprietary to a certain environment. At the same time, the standards may provide connectivity to protocols that are proprietary, or at least are not open.[8] An example of this is the ability of an RFC 1490 implementation to support subarea SNA connectivity. Although subarea SNA is not open, RFC 1490 is because it is not based on a proprietary protocol, nor is the RFC itself proprietary.

8.3.1 RFC 1356

This standard allows for the encapsulation of protocols so that they can be transported across an X.25 network. The method of providing this interface is to define identifiers for some well-known protocols. These identifiers allow the accompanying data to be categorized and correctly processed. This allows a great deal of flexibility in operation and enables many protocols to be passed across an X.25 network.

The X.25 foundation merely provides the road upon which the other protocols are passed transparently as data. The other protocols do not actually operate across the X.25 network. Instead, they are transported as data through the X.25 network. Thus, the X.25 network becomes invisible to the protocols that are passing through.

If some type of session is moving through the X.25 network, the RFC defines only a transparent vehicle to transport the data. The session has no awareness of the X.25 network transportation.

There are two basic methods of data encoding to allow transportation across the network. The first, and more common method, is on a virtual circuit (VC) basis. In this case, a VC is dedicated to a particu-

---

[8]Openness in this context means that the protocol is fully architected and that the architecture is available to the public.

lar protocol. If the VC is a switched virtual circuit (SVC), the SVC is dedicated to a particular protocol during the life of the VC. Once the SVC is torn down, it can then be dedicated to a different protocol when it is reestablished.

The way that the protocol is identified is by placing the well-known protocol identifier into the first byte of the Call User Data (CUD) field of the Call Request packet.[9] This identifier is known as the network layer protocol identifier (NLPID).

Upon receiving the Call Request packet, the receiver extracts this identifier field and is able to determine what protocol will be passed across the VC.

Figure 8.8 shows an example of this process. Note that the CUD in this example specified a protocol identifier of x´CC´. This identifies the enclosed data as being IP data. The sender and receiver are able to identify the encapsulated data and are able to pass IP data across the X.25 network.

The second method is use of the *null encapsulation*. In this case, the CUD use is x´00´. Each message must start with a nonnull NLPID which allows for the determination on a *message basis* of what type of data is encapsulated. This method allows multiple protocols to traverse through a single VC. This can be quite useful if only a limited number of VCs can be supported.

If a VC is accepted that contains the null encapsulation NLPID, it is the receiver's option to discard messages that specify a protocol that the receiver does not support. Because no indication is provided that a message has been discarded, it is useful to determine whether the receiving station can support a particular protocol before it is transported across the X.25 network.

[9]I will not be providing detailed explanations of the process of using X.25. If this is needed, I would suggest that you investigate this topic in more detail.

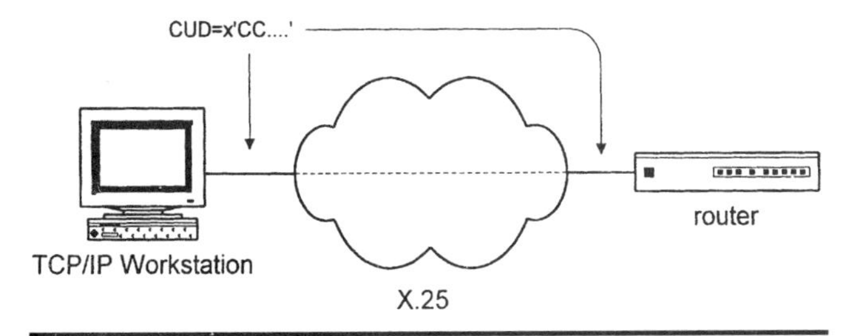

Figure 8.8 IP encapsulation across X.25.

8.3.2 RFC 1490

Frame relay is a DLC that has gained a great deal of attention for data communication. It is a close relative to X.25, in that data is packetized upon entry to the network and is reconstituted at the exit points. Similar to X.25, packets are passed through the network upon virtual circuits that are established when the circuit is requested.

The major differential between X.25 and frame relay is that there is no error recovery *within* the frame relay network. This is based on the fact that data circuits are now of high quality. As such, the overhead of link recovery is exchanged for recovery at the network endpoints. Because errors are sufficiently infrequent, a retransmission through the entire network is rare enough that the cost, in more packets being transmitted across the network, is minimal when compared to the overhead of allowing error recovery to proceed at the link level.

Frame relay carries data across separate VCs. Because each of these VCs is largely distinct from the others, it is possible to carry different protocols across a frame relay network. In a manner similar to RFC 1356, the protocol data units (PDUs)[10] are carried within the user portion of data packets. The protocol being transported is not emulated (or actually present) across the frame relay network; the data is passed without regard to what is within the frame relay packets.

This mode of operation allows entirely different data to be present on different, or even the same, virtual circuits. Thus, subarea SNA, APPN, HPR, and TCP/IP traffic can be passed across a frame relay network by providing a separate virtual circuit for each data type. This is illustrated in Fig. 8.9. The separate VCs allow for the transmission of these different protocols across a common backbone media.

RFC 1490 defines a method of using frame relay VCs to separate different traffic. Each VC is identified through the use of a data link connection identifier (DLCI).

RFC 1490 uses several fields to provide a method of making this determination. The first of these is the NLPID, as in RFC 1356. This field is administered by ISO and International Consultative Committee of Telephony and Telegraphy (CCITT) and provides a standard method of dynamically determining what type of data is being carried within a VC.[11] Through the use of the NLPID, the receiving station can determine what protocol is being carried within

---

[10]A PDU is the user data, plus some original headers, that is transported across the network.

[11]*Information technology—Telecommunications and Information Exchange between systems—Protocol Identification in the Network Layer,* ISO/ICE TR 9577: 1990 (E) 1990 1990-10-15.

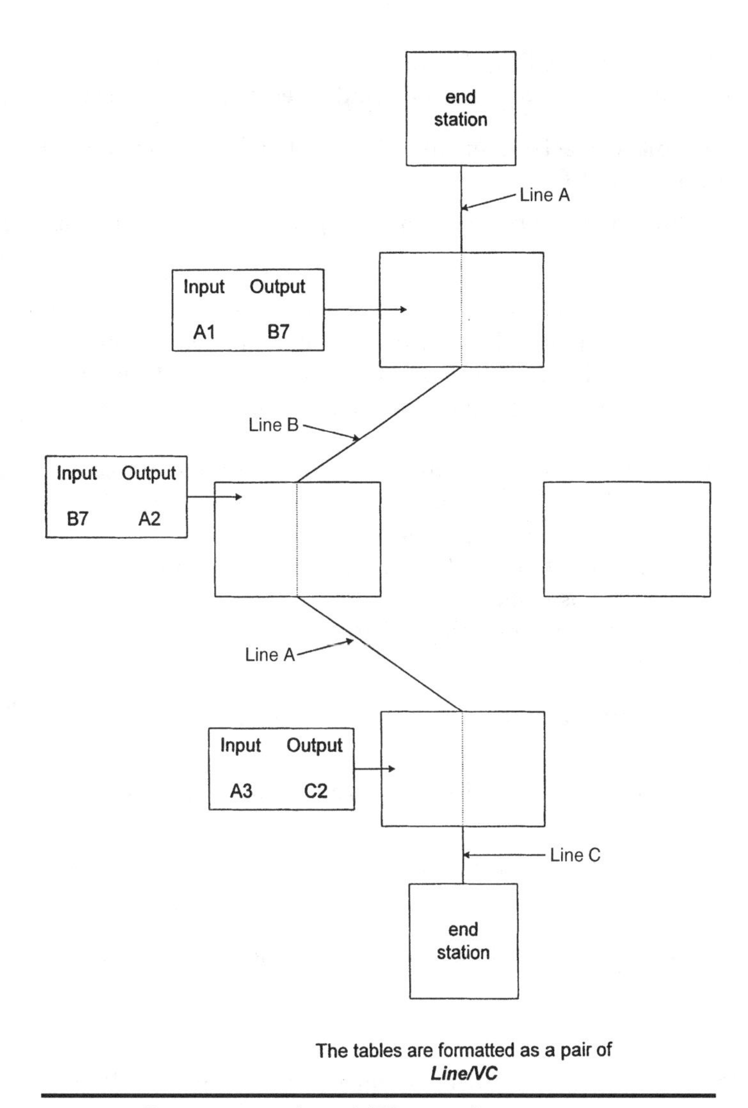

Figure 8.9 Data transport through FR network.

a VC. This allows the receiving station to determine whether that protocol is supported and to make the appropriate adjustments to provide a good interface for the particular protocol.

RFC 1490 also provides a method of protocol identification for those protocols that do not have a NLPID assigned. In this case, the frame contains a subnetwork access protocol (SNAP) header. This header allows for a organizationally unique identifier (OUI) that provides this function. The OUI allows for the definition of a trailing protocol identifier (PID) that fully defines the data that is being carried.

The result of this protocol definition is that a receiving station can:

- Dynamically and correctly identify what protocol is being carried within a frame relay VC
- Determine whether that protocol can be supported and how to support it

Because the frame relay network can provide a meshed network without requiring additional physical connections between each pair of resources; the use of RFC 1490 allows for resources to fully communicate without requiring extensive coordination and capital investment.

One of the major problems with this solution for interconnection is that it is only applicable to a specific backbone environment: a frame relay network. If frame relay is not the selected transport, this solution cannot be utilized.[12] Obviously, this is a major obstacle to the market penetration of this solution.

At the same time, this frame relay solution provides many advantages. These include:

- The ability to support diverse protocols upon a single physical interface
- A standard means to determine the type of data that is being received
- An interconnection method that can be utilized easily by different vendors

(For more information on RFC 1490 and its implications, see Chap. 10.)

8.4 Summary

Several methods exist for integrating subarea SNA, APPN, HPR, and IP protocols into a common network. These methods can be broken down into two types. These are *hardware* and *software* solutions.

Hardware solutions can be used to provide either additional physical paths for the data, or some multiplexing method of passing the different datastreams across a common media.

Additional physical interfaces can provide a feasible solution to

---

[12]The rules established in RFC 1490 can be applied to other environments, but this would fall outside of the intention of this method.

passing diverse protocols across a network. The simplest way is allowing for additional physical interfaces and lines can solve the problem. The protocols are separated physically to allow them to be transported. Though there is no actual integration of the protocols into a common base, it is possible that management of the network may be done from a common platform.

There may be valid reasons to take this direction for integration. For example, if the additional protocol only has to be supported for a short period of time, it may not be economically viable to create a physically integrated network. In this case, adding hardware interfaces for the short-term is the correct solution.

It also may be desirable to use separate physical paths if there are extenuating circumstances. For example, physical separation may be preferred because of security requirements.

There are hardware methods of multiplexing multiple, diverse datastreams into a common base. The protocols are actually integrated into a common foundation that allows the transportation of diverse protocols. Though a pair of multiplexers often use a common proprietary protocol between them, this physical path can carry data from multiple sources and destinations.

Software techniques also exist for the transportation of different protocols. The simplest of these provides for the same type of multiplexing as is exhibited by the hardware multiplexers. The distinguishing factor of the software solutions is that they are easier to upgrade and that they are normally lower in cost. At the same time, they are generally slower and require a platform upon which they must operate.

Software solutions also include several "standards-based" methods that allow diverse data streams to be incorporated into a common base of data transport. These standards are largely a product of the IETF and are contained within several RFCs. Among these solutions are RFC 1356 and RFC 1490.

Both of these solutions define a method of encapsulating the supported protocols onto a common transport base. They also standardize how the messages that are transported can provide self-definition of what protocol is contained within the packets. This self-definition is used by the receiving station to determine how to support the incoming data. This support may be done by not supporting the protocol. In this case, the receiving station would reject either a connection that specified a protocol that it did not support, or would discard packets that specified one of these protocols.

If the protocol is supported, the receiving station can provide the necessary support needed to pass the encapsulated data to its destination and the protocol support necessary to perform this transport.

These methods of passing data from diverse sources allow companies to create networks that can transport information between systems that do not normally allow communication. They also allow these same companies to optimize their network by providing a common transport for different protocols.

Chapter

9

Protocol Transport and Conversion

Chapter 8 discussed several methods to provide an integrated network transport for a diverse set of communication protocols. This chapter takes a closer look at the conversion of a communication protocol in an effort to create a common network protocol throughout the backbone network.

There are two methods of providing this type of transport. The first is to implement a communication transport that passes the nonnative protocol through the native transport. In this case, the entire nonnative transport frame is encapsulated within a native transport frame. This provides a transparent transport for nonnative protocols. This option is available from many vendors including cisco, in its SNA Tunneling (STUN) product, and Hypercom, in its transparent transport product.

The other option is to use protocol conversion to allow the network to isolate noncommon protocols to the periphery of the network. Fig. 9.1 shows how a network would appear when using this method. All of the noncommon protocols are translated at the periphery to the common protocol. This creates a homogeneous backbone network that is clear of the issues associated with heterogeneity.

On the other hand, the conversions/translations that are done at the periphery must be a high quality to provide a transparent and effortless movement throughout the network. It is also strongly advised that network management be able not only to access the boundary of the network, but also to pass through that boundary to access the actual device.[1]

[1]There are *few* conversions that provide this and none that provide this support for all interfaces. The normal level is to support an emulated management interface at the boundary that depicts the condition of the emulated device.

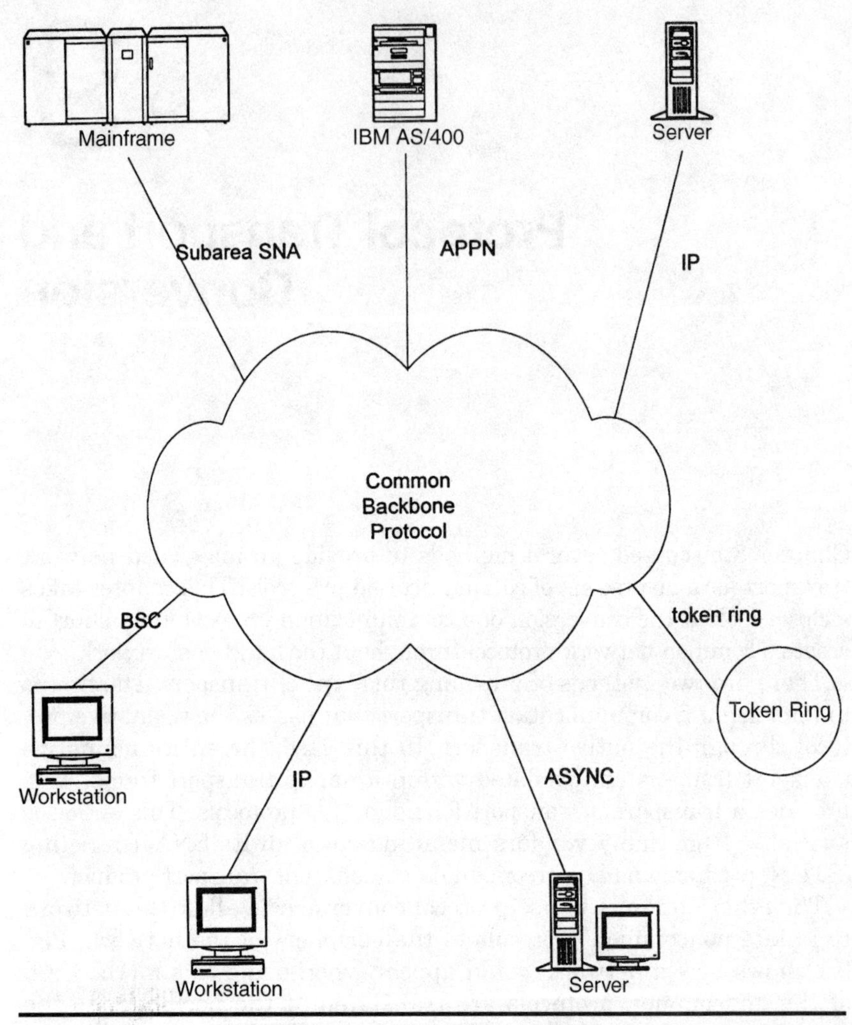

Figure 9.1 Noncommon protocols on periphery of network.

Figure 9.2 illustrates the distinction between these two options. As you can see, the ability to interrogate the actual device is strongly desirable! Without this ability, the network control staff are at the mercy of the conversion interface to accurately reflect the status of the remote resource.

Several types of conversion can be provided. The distinction is the completeness of the conversion. In some cases, the conversion provides a terminated protocol. This type of conversion provides some

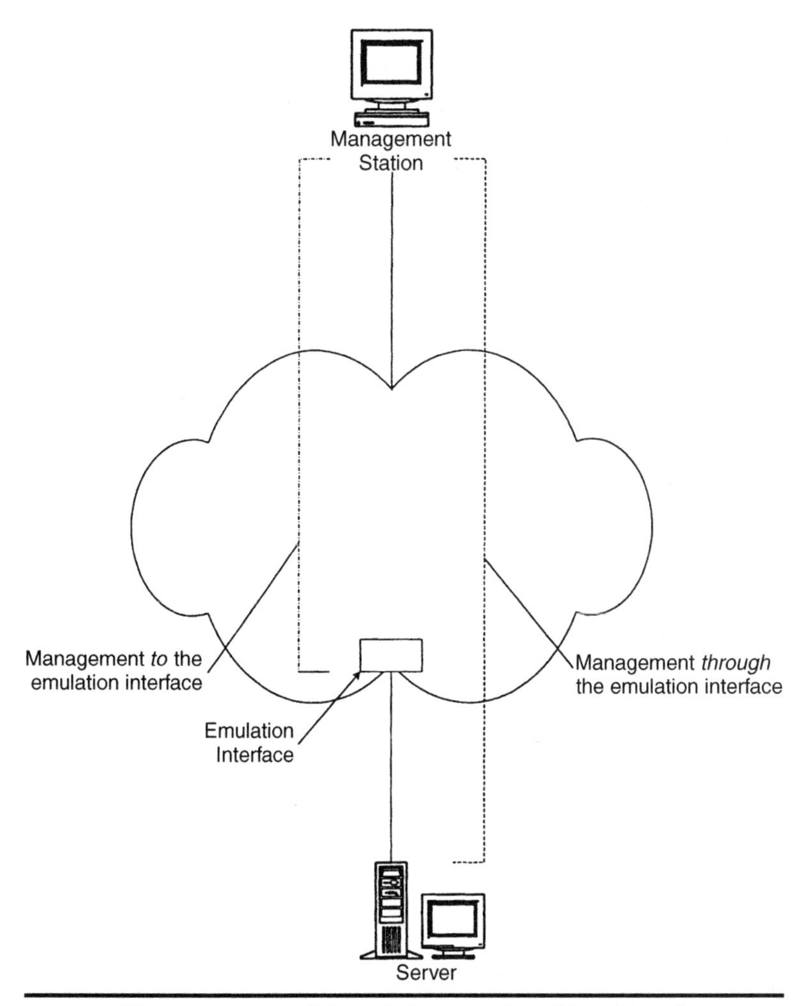

Figure 9.2 Management to and through the network peripheral interface.

minimal aspects of the emulated protocol that allows data to be transported in a minimalistic manner. This type allows user data to be pulled out of the noncommon protocol and placed into a frame of the common protocol. Any address translation is done through a simple table lookup method and network management is limited to the status of the interface, rather than the actual device.

At the other end of the spectrum, there are conversions that can provide a clear connectivity path between the noncommon protocol and the protocol within the backbone. These conversions often use algorithmic methods of address translation to reduce the definition requirements for network operation and support. These conversions may also allow for dynamic network topology changes. These changes can be adapted to without losing connectivity to the network endpoints.[2] Some conversions even allow you to disconnect the physical network topology from the logical configuration. These conversions can optionally allow you to logically reconfigure your network without requiring a physical change.

An example of this is implemented by the Hypercom Integrated Enterprise Network (IEN) products. This architecture supports the logical reconfiguration of the network into topologies that do not actually exist. It also allows for the logical coupling of resources that are connected to noncontiguous connection points. Figure 9.3 shows this type of connection.

9.1 Transport Encapsulation

The simplest method of passing nonnative protocol through a backbone network is to encapsulate the nonnative protocol within a native transport frame. This allows the nonnative requests to pass through the native backbone transport unimpeded and without modification. The frame is then transmitted to the nonnative interface as it normally would be seen.

By creating a transparent transport through the native network, the nonnative protocol is merely transported without any emulation or conversion. If done perfectly, the nonnative interface has no awareness of the intermediate network transport.

Figure 9.4 shows this type of transport. In this case, the nonnative protocol (SNA) is transported across the native, backbone network (frame relay) without any emulation or conversion. All frames are passed between the nonnative nodes without any modification.

The inherent problem associated with this type of interface is that some protocols have constraints on the timing of the interface. These constraints normally have timeouts for responses and other timer limitations. If a response is not received within the specified time limit, the interface is disabled with a timeout.

[2]This is a strongly desirable aspect of these conversions, as the network topology is learned dynamically. In addition, these conversions are able to adapt to changes in the network without operational changes.

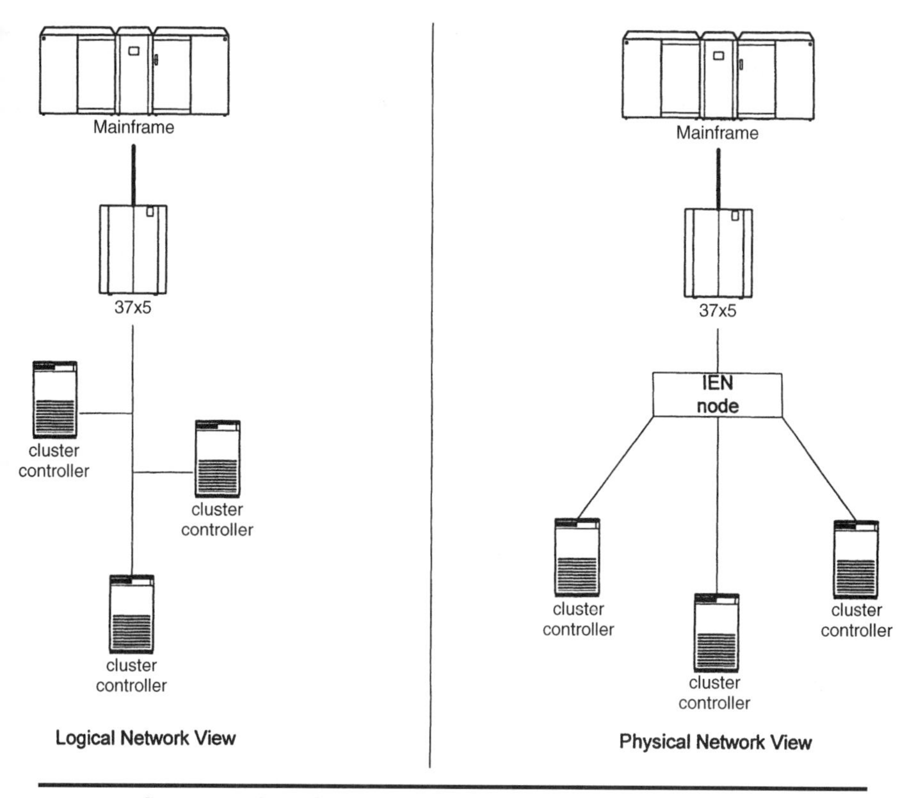

Figure 9.3 Contrasting logical and physical network views.

Because of the overhead of the associated native transport, it is not abnormal for a response to not be received within the normal time limit. Sometimes these timeouts can be mitigated by modifying the normal timers to allow for the longer response time.

Another problem can be seen if the backbone must transport such link-level messages as SNRM, UA, and RR. In this case, the backbone network can become bogged down by the overhead associated with these link-level messages.

This type of problem can be overcome by eliminating the RR from the backbone network. Since the RR message normally is transmitted many times per second, the elimination of these messages can relieve the backbone. These messages can be eliminated from the backbone by *spoofing* them at the point of encapsulation. Thus, Interface A and Interface B can spoof the RR at their respective local interface. This offloads the backbone from transporting these link-level messages.

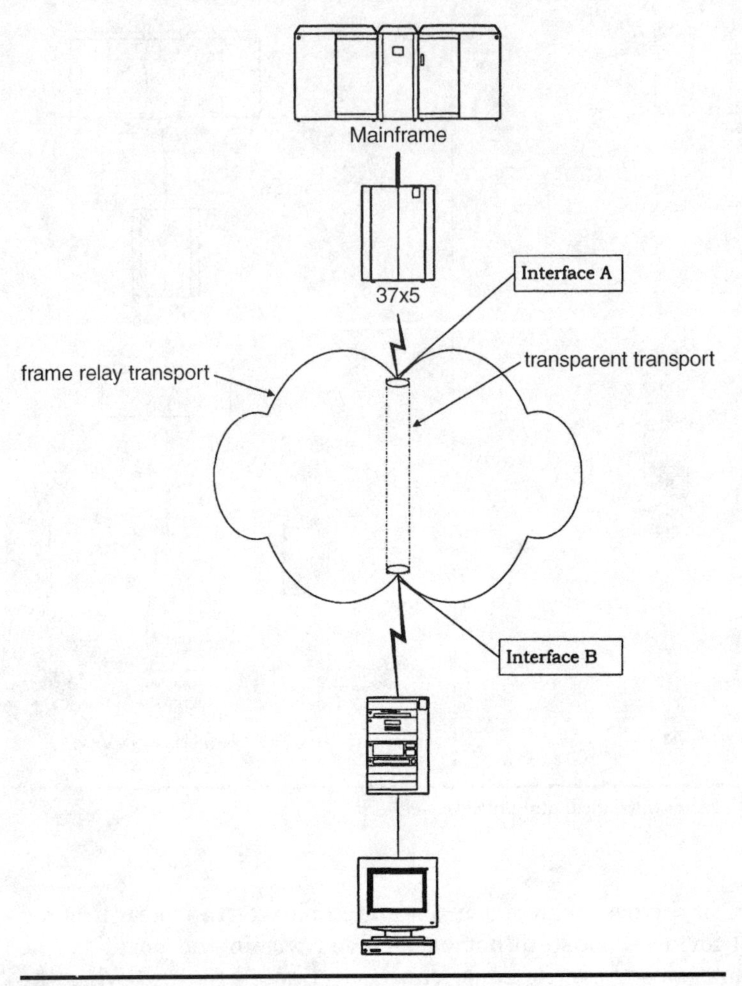

Figure 9.4 Transparent transport of nonnative protocol.

9.2 Protocol Conversion

The purpose of protocol conversion is to provide a gateway-type product that can terminate one protocol and originate a second one. This is a highly desirable product, if the conversion can be well done.

The conversion interface has a dual image. It creates a gateway component at the boundary of the common-transport network. This gateway image is provided by existing within both the common-transport network and the external noncommon portion of the larger net-

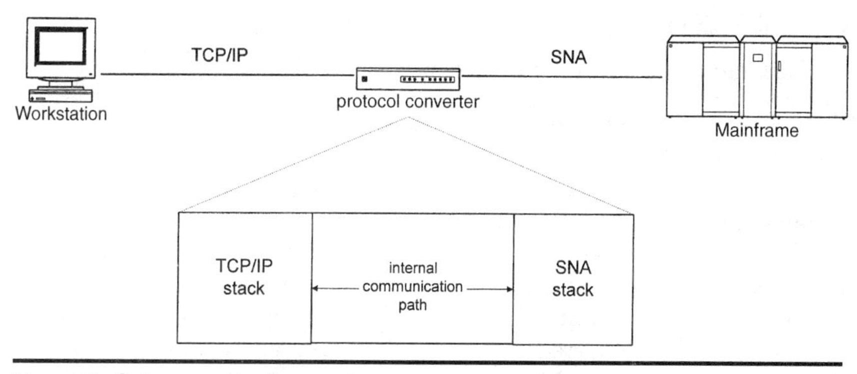

Figure 9.5 Gateway protocol converter.

work. By existing in both parts of the network concurrently, it has the ability to see into both sides. This provides it a unique spot in the network by supporting a view of the entire network.

Figure 9.5 shows how the gateway product is situated within the network. You can see from this figure that the dual image of the conversion product affords it the ability to transfer data across the boundary between the network portions.

9.2.1 Crossing a protocol boundary

Terminating a protocol is a far more complicated activity than it may appear at first glance (and at first glance, it does not look easy!). An excellent understanding of the protocol that is being terminated is necessary.

In the case of the four protocols addressed by this book (subarea SNA, APPN, HPR, and TCP/IP), they all have in common the fact that they are connection-oriented protocols. That is, they all establish a session between endpoints in the network. The purpose of this session is to provide the view that the endpoints are directly communicating.

Each endpoint has a level of control over the communication between them. This control includes the ability to:

- Establish a session
- Terminate a session

- Control the flow of data between the session partners
- Pass signals to the session partner

As a result of these controls, the session partners are able to determine the status of their partner. These controls also allow partners to ensure that data has been successfully passed to the session partner. This is one of the key advantages of using a connection-oriented protocol. Data integrity is kept at a high level by providing almost immediate feedback on the status of communications.

One area of communication uncertainty is the size of the *transmit window* that is supported by the session partners. All of these protocols support different-sized transmit windows, but a modulo 8 window is typical. In this case, no more than eight frames can be outstanding before an acknowledgment is required. This both allows communication to proceed efficiently, by allowing multiple messages to be transmitted without an acknowledgment, and limits the size of the uncertainty within the network on the status of messages.

Knowledge of the window is also important to the session receiver. If the receiver is expecting a window of 1, but the sender is operating with a window of 8, the receiver will not be able to respond to all of the incoming messages. It is probable that the receiver has allocated only one receive buffer. In this case, the receiver will not have sufficient receive buffers to handle the eight incoming frames.

These protocols also use a "keep alive" message that ensures the session endpoints that its partner is still operable. This message is transmitted when no user data is being transmitted for a period of time.

The SNA protocols are *polled* protocols. These protocols use a primary/secondary dichotomy between the session partners. The secondary partner cannot transmit until the primary partner polls the secondary. This can cause problems if the data flow is largely unidirectional toward the primary partner, but the polling is normally done at a sufficiently high rate to reduce the likelihood of bottlenecks.

TCP/IP is not a polled protocol. It uses a full-duplex session interface to pass user data between session partners. Use of a full-duplex interface results in a session partner being able to transmit data whenever the transmit window is open.[3]

Because of the differential between the SNA and TCP/IP protocols, it is possible for a bottleneck to start to develop at the SNA interface

[3]A transmit window is said to be *open* whenever the transmitter is able to send data.

as TCP/IP frames arrive, but cannot be transmitted to the SNA session partner until it is polled.

The problem is not large, if the transmit windows are kept the same. In this case, the TCP/IP acknowledgment should be withheld until the frames can be transmitted on to the SNA session. Only after the frames are transmitted, should an acknowledgment be sent on the TCP/IP session. By operating in this manner, there are never more frames outstanding than fit within one window.

If the TCP/IP session is at the primary end of the SNA session, greater control over the throughput is possible, as it is the primary SNA session partner that controls polling. In this case, the protocol converter can regulate the transmission of data on both the TCP/IP and SNA sessions (as it could in the case of the SNA secondary), but can also provide greater control over the reception of data on the SNA session by being the one that polls for data.

This control, though, is quite limited. This is because the primary session partner cannot decide that it cannot tolerate receiving more data, so it just does not poll. As I said, SNA sessions are quite sensitive to various parameters on the session. One of these is the secondary timing of being polled. If the secondary session partner is not polled for a certain period of time, it assumes that the primary partner has failed or that the session has failed because of a lack of communication. As the loss of the session is not one of our desired outcomes, this is not a viable solution to the problem of throughput.

It is possible for either session partner to send a signal that no more data can be transmitted. This indication is sent as a Receiver Not Ready (RNR). As the name implies, this is sent when the receiver cannot tolerate the reception of more data.

On receiving an RNR, the receiving session partner must terminate the transmission of user data frames until a Receiver Ready (RR) is received. Figure 9.6 illustrates the use of the RNR and RR command indications. Note that this effectively controls the transmission of frames on an SNA session.

9.2.2 Maintaining session status across the protocol boundary

As all of the addressed protocols are session oriented, it is possible to maintain session status across the protocol boundary. Though two sessions are maintained, it is possible to coordinate session status across the boundary between the sessions. Thus, the discontinuous sessions can be managed as if they were a single session.

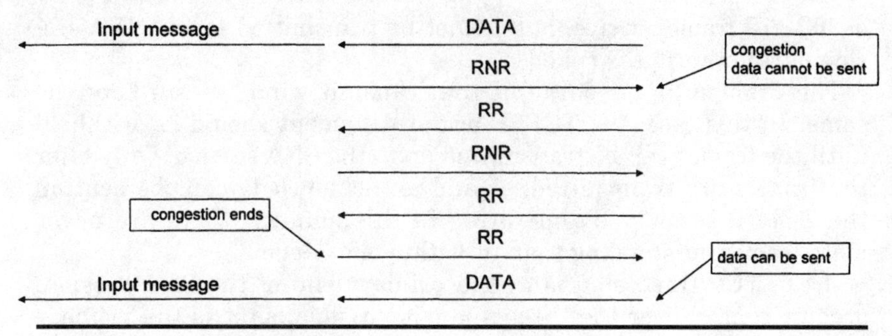

Figure 9.6 RR and RNR processing.

If one of the sessions is lost, it is possible to signal the partner session. This signal would notify the other session to also terminate.[4] This would maintain the "illusion" of a contiguous session.

9.2.3 Subarea SNA and APPN

Though this is not normally thought of as requiring a protocol converter, it is an interesting area to investigate for the use of this type of device.

APPN was developed to address some of the weaknesses of subarea SNA. At the same time, it was not designed with a method of interacting with the subarea SNA architecture. As a result of this, there are many areas in which the architectures do not mesh well.

One of the key design requirements for VTAM V4R1 was to address the migratory issues raised by attempting to integrate APPN into a subarea SNA network. In some areas, such as session services, the level of code required was considerable in order to provide the smooth migration demanded by IBM customers.

Because APPN is an open architecture,[5] several companies have developed APPN access. Though IBM still keeps some of the "family jewels" closed, other vendors have started designing their own extensions to APPN; in some cases, these extensions have been brought to the AIW for discussion.

---

[4]Proper management of this session could be maintained by establishing a timer that determines how long the first session must be down before the second session is terminated.

[5]The openness can be seen by the fact that IBM has conducted the APPN Implementors Workshop and provides architecture for several components, including the development of DLUr/s, HPR, and DLSw.

It is feasible for a vendor to develop a value-added product for use in this arena. Some areas that could be useful include:

- Extended network management from that supported by VTAM and Netview
- Extended connectivity, such as MLTG
- Allowing cross-network connectivity without implementing extended border node architecture

The VTAM/NCP composite node was developed to address the interaction of subarea SNA and APPN. This node type has this task as its reason for existence and it does a very effective job of providing this support.

It is only available when both VTAM and NCP are being utilized. It does not address the APPN customers that do not use these products, but have extensive APPN networks. An example of this is an AS/400 network. These networks use APPN as their native network protocol. Yet, the support for subarea SNA devices is limited and difficult.

A product that can support the older subarea SNA devices while allowing connectivity to the APPN backbone network can be very useful. This type of device can be especially important when APPN networks must be merged. Because VTAM V4R2 is the only APPN product that has implemented extended border node architecture, a conversion device could be useful if it allowed for the segregation of networks, while concurrently supporting sessions between those networks.

Though this product may not use either extended or peripheral border node architecture, this does not preclude the converter from providing a similar type of support. It would supply this support in the same manner that protocol converters are able to connect subarea SNA and TCP/IP networks; one protocol is terminated and the other is established within the converter.

An APPN/subarea SNA converter would provide the same type of service for the devices within each network. If the converter terminated a session in one network, created a private communication method across a network boundary, and established a session in the other network, the end result would be similar to the extended border node architecture, except that it would not follow the proscribed architecture.

At the same time, since border node architecture is being opened to the AIW, there is nothing to preclude a vendor from proposing an alternative method of providing communication across a network boundary. Though this may not be the method used by IBM in VTAM, it is perfectly feasible that this new method may find a following in

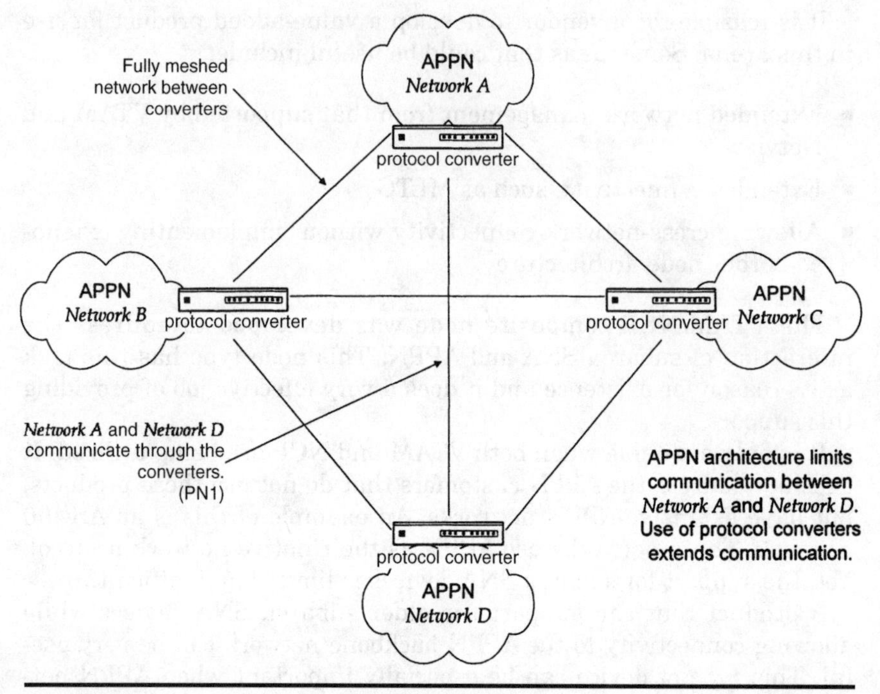

Figure 9.7 Use of protocol converters to allow communication.

the vendor community and define the architecturally approved method. IBM is then stuck in the interesting position of deciding if they want to use the method approved by the AIW.[6]

Figure 9.7 shows how this design could operate. In this case, the termination of the session in Network A allows data to be passed between Network A and Network D by crossing the private network, shown as PN1. Though the networks are not connected architecturally, data can freely pass between the networks. In addition, it is possible to pass network management requests between the networks. By using a private service point interface on the converter, it is possible to depict the entire network topology, including the network names, TGs, and the path between the networks to NetView.

Because Network A and Network D are not adjacent, normal architecture dictates that sessions cannot be established, unless an extend-

[6]Because the border node falls outside of the APPN licenses, this is a perfectly valid scenario. This is quite similar to what occurred with RFC 1434 (DLSw).

ed border node is implemented. Yet, sessions (or the appearance of a session) can be established between resources in these networks.

The protocol converter creates the illusion of an end-to-end session. Though this actually consists of two sessions, with a private data transport between the networks, communication between the networks is complete. The logical fact that this is done by passing data between two sessions does not diminish the fact that data can freely pass between the networks. Users, which do not care about architecture, but instead about "getting the job done," have the communication that they want without the requirement for VTAM V4R2 or NCP V7R2.

Since various vendors, including IBM, have been using this discontinuous session architecture for many years, this is a method that has been implemented successfully many times. It is also a method that is appropriate for this purpose.

Such products as IBM's NRF use exactly this type of dual-session architecture. As NRF is used by customers passing such sensitive data as financial information, this method is a fully supported and utilized method.

9.2.4 SNA and TCP/IP

SNA sessions can be terminated and the data placed into a TCP session. In this case, the SNA session[7] is terminated, as was described in Sec. 9.2.3. The important issue is that the session layer be correctly emulated so that the SNA session is faithfully implemented.

With the SNA session terminated, the user data can be extracted and moved into a TCP session. The TCP session, as is true of the SNA session, must be correctly and fully implemented in order to support the transportation of the user data.

It is now the responsibility of the protocol converter to move the user data between the sessions. It is also the converter's responsibility to respond correctly to any control flows on either session. These flows include:

- Bracket initiate and termination (SNA)
- Change of direction indication (SNA)
- ACK on the TCP session
- TCP sequence numbering

---

[7]It is immaterial whether a subarea SNA or APPN/LU 6.2 session is being terminated.

By viewing the two sessions as being discontinuous, it is possible to provide an effective method of passing data between these session types. The user data is easily moved between the session, as long as the session flows are correctly implemented.

Figure 9.8 shows how this is done. Note that the SNA session and TCP sessions are managed independently. All control flows are implemented individually for each session in order to provide correct implementation of the necessary protocols. By operating independently, the flows for each session can be optimized, without regard to impacting the other session or session types.

By using this type of protocol converter, it is possible to create a homogeneous backbone network that has connections to nonnative interfaces only at the periphery. This eases the task of creating and managing the backbone network, in that there is a single protocol evident in that backbone.

9.2.5 Protocols reversed

It should be understood that the protocol converter setups that have been described are not exclusive; the reverse set of protocols is also a valid configuration. Thus, it is possible to convert SNA protocols to TCP/IP, for transport across an IP backbone. In this case, the SNA sessions are along the periphery of the network. The protocol converter, in this case, handles the management of the SNA session layer to provide connectivity to the SNA resource on the outside of the backbone network.

In a similar fashion, it is possible to handle connection to an APPN network along the outside of the backbone network that is operating as SNA subarea. The use of this type of connectivity allows resources in the APPN network to communicate with resources inside the subarea SNA network, without requiring use of a composite node nor that the APPN network connect as a LEN node.[8]

9.3 Network Management

Network management is a key requirement if protocol converters are to be utilized. Without management of both the converters and the links on both sides of those converters, the network can fall into dis-

---

[8]Though it is possible that LEN connectivity is not supported by the VTAM level that is operating, since this support goes back considerably, it is assumed that this support exists.

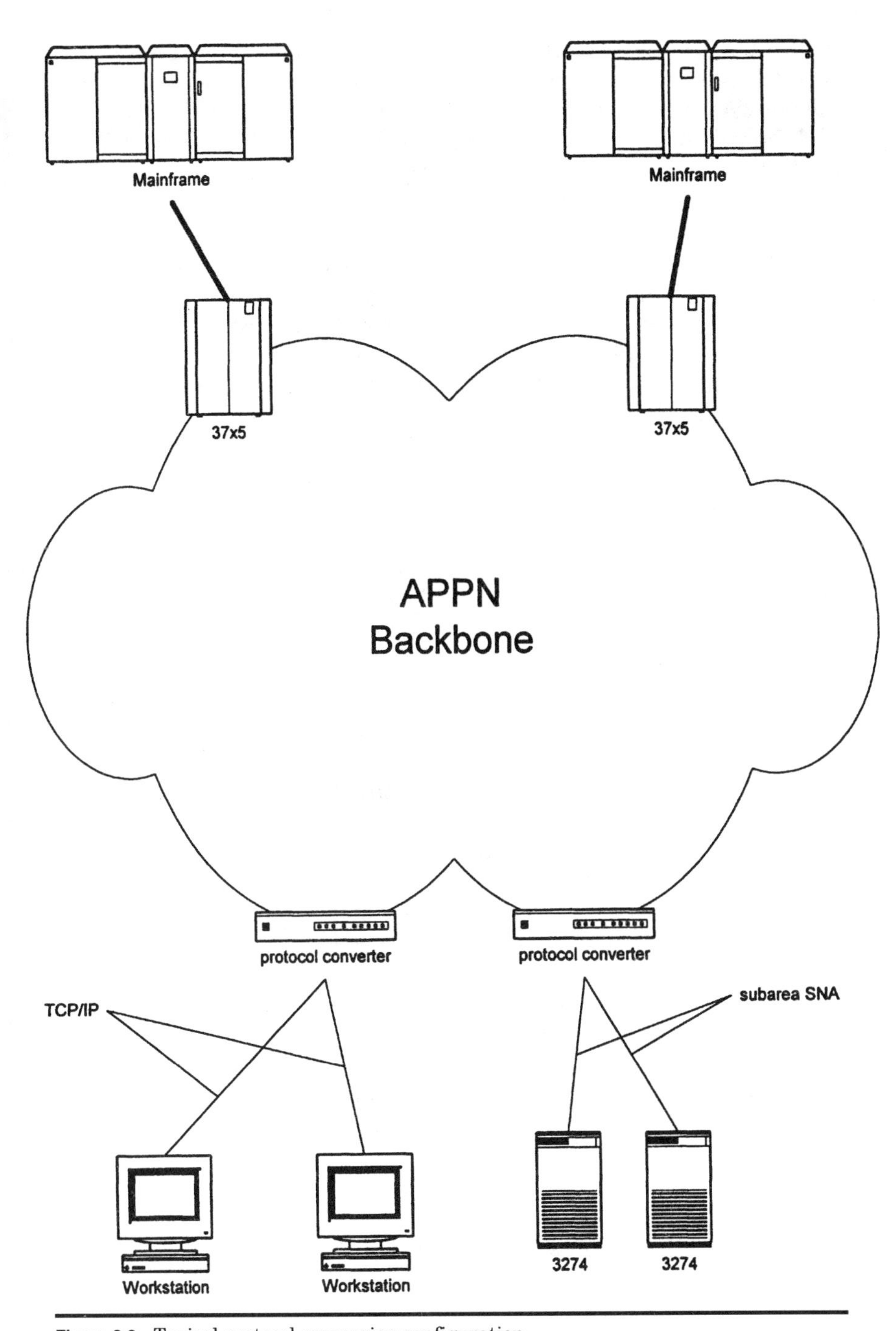

Figure 9.8 Typical protocol conversion configuration.

repair. The cause of this is that without this level of management, the network can suffer failures without anyone's knowledge. As a result, the network becomes unavailable for use.

If the purpose of the protocol converter is to enable use of the network, lack of management is directly counter to this aim. All networks must be managed; the fact that a protocol converter is in use does not reduce this requirement.

At the same time, the use of a protocol converter can make the task of management more difficult. This is a result of the fact that the converter "hides" the portion of the network on the outside of the converter from management on the inside. What is required is that the protocol converter cooperate in the management task by providing a window to the outside network.

9.3.1 Management to the conversion interface

Network management can stop at the network boundary interface. This allows management to be conducted only up to the network boundary at the protocol converter. This management function allows the internal backbone network to be managed, and can determine if the protocol converter is active and available. Information on the data throughput and detection of errors outside of the backbone is not supported.

This limited management interface provides information only to the protocol converter itself. There is a "black hole" on the other side of the converter that cannot be penetrated for management purposes.

Information on the protocol converter normally includes very limited information that can only be inferred by indirect information. For example if the converter operates by converting SNA to TCP/IP, if the TCP session that terminates at the converter shows an active state, you can reasonably assume that the protocol converter is active.[9] This configuration can be seen in Fig. 9.9.

Though some network management information is available, the details are limited. Because of this limitation, this configuration is discouraged.

---

[9]Because of the long timeout of a TCP session, this may not be an accurate depiction of the state of the protocol converter, but this is the best that can be derived from the limited information.

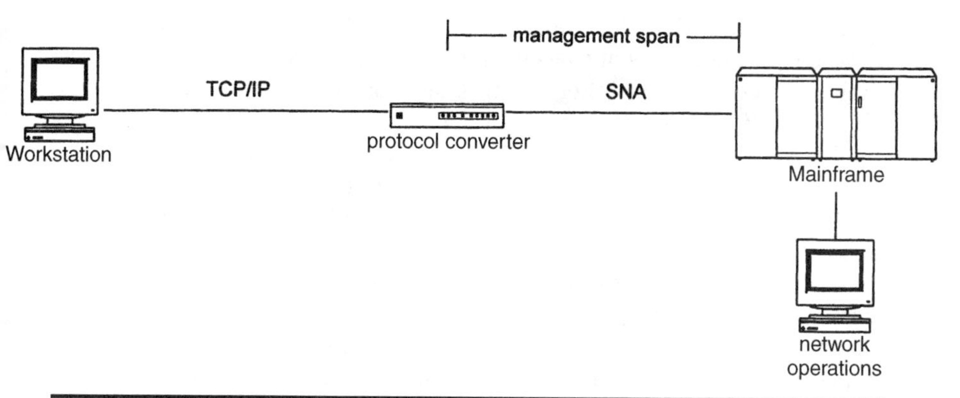

Figure 9.9 Management to protocol converter.

9.3.2 Management through the conversion interface

Management can be extended to provide knowledge of the communication path that is on the noncommon communication interface. This view of the external network can be done by gaining knowledge of the communication on the other side of the protocol converter. This management span can be seen in Fig. 9.10.

At a minimum, this management window should provide a query function that allows the status of the external network to be investigat-

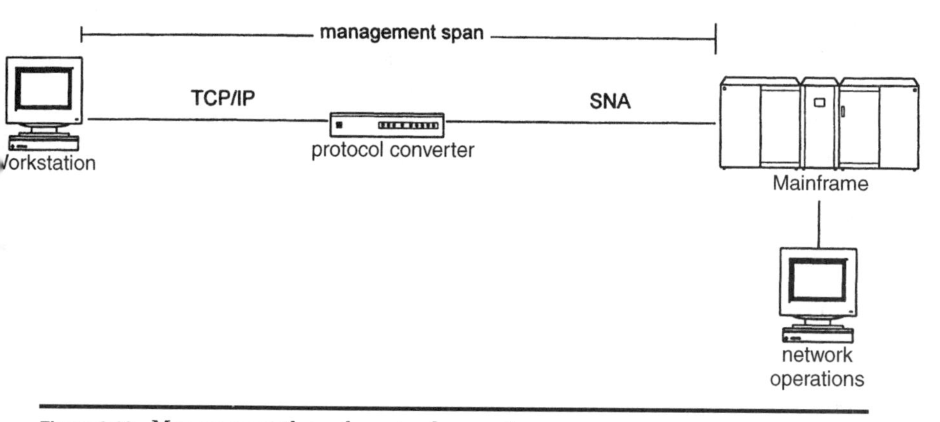

Figure 9.10 Management through protocol converter.

ed. This would allow the outside network to be queried for status. If a status is found that is of concern, it can be further investigated and possible corrective action taken. At this level of management, the network control staff can at least determine, proactively, if the outside network is available and whether corrective action is needed. There is not a need to operate in a reactive fashion to user statements of network failure.

The next level of management would provide asynchronous alerts of a network event, such as a component failure. This mode of operation allows the network operation staff to become aware of a failure without having to inquire about the current status of a resource. This greatly eases the task of network management, reduces the time until a failure is detected, reduces network workload by reducing the amount of idle inquiry traffic, and potentially reduces the duration of an outage by allowing quicker detection.

Another level of management is the allowance for the converter to not only allow inquiries, but also to control the outside network. This provides the ability of a centralized network management group to control the backbone and peripheral network components. In addition, this control can be done from a location that is neither local to the outside network nor has direct knowledge of this external network.

By having the protocol converter provide the interface into the external network, it is possible to determine the status of the external network as seen from the converter. Providing a control function to the outside network facilitates a unified view of the network, even though the components may not actually be integrated. This view can be brought closer together by allowing cross-reference of information between the interior and exterior networks. Forms of this correlation can be of use when viewing nodes that provide the gateway to and from the exterior network. These nodes can provide a dual image that allows the network boundary to be viewed and analyzed.

The top level of network management occurs when the protocol converter not only allows the network boundary to be interrogated and controlled, but also includes interfaces for tools to be built and analysis conducted. These network management tools can facilitate greater control over the network boundary by allowing more information to be collected. Such items as controlling network boundary throughput, automated adaptation to network events, and tools to allow more network analysis to be conducted can be included in the network operation.

9.4 Summary

Transportation of a nonnative protocol can be provided by either creating a transparent transport through the native backbone network or by converting the protocol into that of the backbone network. In both cases, the data is transported to the destination across the common backbone network.

The creation of a transparent transport through the backbone allows any protocol to be transported across the backbone. Since the entire transport frame is transmitted as user data, the backbone network is not aware of the type of data being transported, which allows the message to be passed, nor are the endpoints aware of the intermediary transport network.

Ideally, this solution provides a simple method of passing any type of protocol through the backbone network. In reality, problems arise. These problems center around the problem of adding the overhead and time associated with passing the transport frames through the backbone network.

All of the protocols discussed in this book (subarea SNA, APPN, HPR, and TCP/IP) have timeouts associated with the lack of a response within a specified time period. These timeout periods are not built to include the additional overhead of an intermediary network. As a result, timeouts can result from the additional network transportation.

It is possible to specify longer timeouts for some of these events, but not all events have the option of a configurable time period. In addition, in this configuration, some link-level messages can be, optionally, *spoofed* at the point of entry to the backbone network. An example of this type of message is an RR for SDLC-based SNA interfaces. The spoofing relieves the backbone of having to transport these idle link-level messages. This both reduces the overhead on the backbone and increases the reliability of the link-level messages to be answered.

Another alternative for connectivity is to convert the nonnative protocol into that of the backbone network. This protocol conversion allows the nonnative protocol to be correctly coordinated at the periphery of the network and also allows a common transport protocol to be contained within the backbone network.

Care must be taken to ensure that the protocol converter is of sufficient quality that it provides a reliable protocol interface. The converter should also allow some interrogation of the nonnative network for network management purposes.

The protocol converter provides a dual image to the two network interfaces. The first image is a termination of the nonnative protocol. This protocol termination creates an endpoint for the nonnative protocol. At this point, the nonnative interface appears to have reached its destination.

A second protocol interface is initiated at the converter. This second interface provides the path to the actual data destination. User data is extracted from the first interface and passed onto the second. The end user's view is that a contiguous path to the data destination exists. Theoretically, these users have no awareness of the discontinuous nature of the communication path.

In order to actually provide this transparency, the protocol converter must adhere to the respective protocols and allow some type of status coordination to be conducted by the protocol converter between the two sessions. By allowing this composite view of the network, the protocol converter can partake in the network management and can accurately reflect the status of the complete communication path. Thus, if a composite view of the network is available, the converter can show a nonactive status if either of the two sessions is unavailable.

As has been alluded to, network management in this configuration can become a challenge. This is caused by the lack of a complete picture of the communication path and its status. In the case of the encapsulated, transparent transport through the backbone network, depiction of the actual transport path is difficult, if not impossible. This is because there must be a method of gaining awareness of the path that the transparent data takes through the backbone network and associating this with the path from the nonnative attachment and the data destination. This is a very difficult association to make and one that can often dynamically change. The ability to depict this path is normally not included. It is the responsibility of the network control technician to understand that this is occurring and to determine whether connectivity exists through the backbone.

Network management of the protocol converter can occur in two modes. The first is a limited management interface that terminates at the converter; there is no awareness of the status of the remote nonnative network by the management interface of the backbone network. Without this remote awareness, it can be difficult to detect a network outage and challenging to determine the point of the outage.

The second mode of operation allows knowledge of the remote, nonnative interface to be communicated to the backbone management platform. In this case, knowledge of the remote network interface is passed to the backbone management platform. This situation allows for an accurate picture of network connectivity to be developed and utilized.

The difficult point for the second mode of operation is to create a way of depicting the accurate status of the remote interface, while concurrently providing a consolidated view of the network availability. Often these are conflicting requirements that result in some type of compromised management view of the network.

Chapter

10

RFC 1490

RFC 1490 was developed in 1993 as a method of encapsulating multiple protocols within a frame relay (FR) network. This method builds upon the base created in RFC 1294 from the year before.

The base purpose of these RFCs is to create a common transport that multiple protocols can traverse. By using an encapsulation scheme, the noncommon protocol frames are enclosed within a frame that can be transported across a FR network. Through the use of RFCn 1490-compliant equipment, the FR network can become the common transport for all of the protocols utilized within an enterprise.

10.1 Fundamentals of a Frame Relay Network

An FR network is not a simple network topology that only provides a transport connection from Point A to Point B. In addition to the reduced overhead of the FR transport, the FR network also adds functionality that has direct implications on how you operate your network.

Among these functions are:

- Dynamic rerouting of traffic to an alternate route
- Virtual point-to-point connection
- Bandwidth on demand
- Discarding of messages in the case of network congestion
- Endpoint error recovery
- Congestion notification

Together, these functions can change the view of where certain functions exist within your network. For example, because the FR network provides a level of dynamic rerouting of traffic, the requirement for other layers of routing may change in your network.

10.1.1 Dynamic rerouting

Frame relay networks automatically redirect traffic to new physical circuits if a loss of connectivity along a path is detected. As a result, the FR itself provides for a certain level of redundant pathing not exhibited by a leased line backbone.

By providing this feature at the link layer, the requirement for an elaborate routing algorithm at the network layer is eliminated. Although routing is needed to determine what FR connection point should be utilized, the need for routing across the interface disappears.

If the backbone network provides dynamic rerouting of traffic, one could question why another routing process is required. Routing protocols are not simple nor do they have a low overhead. If the "network" automatically provides for the rerouting of traffic,[1] what routing decisions are left? The answer is the routes to and from the access point to the FR network itself.

Figure 10.1 shows a typical network using frame relay as the backbone topology. In the case of this network, access to the FR network is local to each location because there is no reason for a higher level of routing. All connection points are a single hop from the destination. Even the backup destination and peer connections are only a single hop away. A higher-level routing protocol provides no additional functionality.

This figure shows a network that requires the routing of traffic to and from the FR access point. In this case, routing decisions need to be made to traverse the network on each side of the FR network. This enables the FR network to appear as a single hop between islands on each side of the network. As a result, routing algorithms will be aware of the adjacent access points on each side of the FR network.

If a subarea SNA network is being used, the network designer defines the link across the FR network as an explicit route between the 37x5 on each side of the network. The session virtual routes are mapped on to this explicit route.

---

[1]As was stated earlier, the FR network appears to consist of point-to-point links; no destination is more than one hop away.

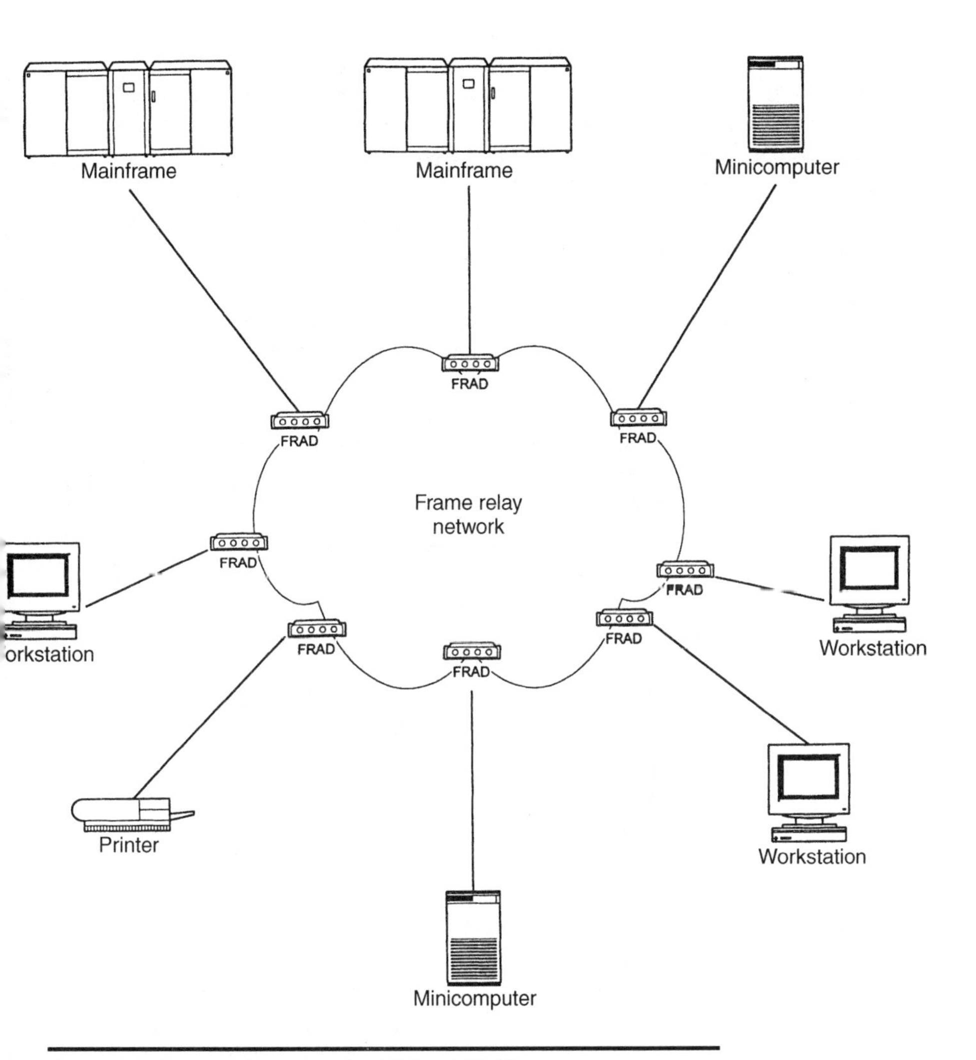

Figure 10.1 Frame relay network with local FRADs.

Decisions about class of service and prioritization can be made by including or excluding VRs that use this explicit route in the definition of the COS table. Sessions that request a definition from the COS table, enables the transport to use it. Other sessions that do not include this route will bypass the FR network.

The optimal method of viewing a FR network in an APPN network is as a connection network. In this case, the FR network is viewed as possessing adjacent links to all nodes that have TGs specifying the same connection network. The normal APPN routing algorithms view this as a large set of adjacent links that can be used effectively and efficiently to allow access.

CP-CP sessions are brought up as required sessions and are defined across the FR network. In addition, LU-LU sessions utilize the FR network, if it meets the session characteristics specified in the COS table. For example, if the FR network is defined as having a low security profile, but a session demands a high security profile, the FR network becomes inappropriate and is discarded. If, alternatively, the hop count is the deciding factor, and the alternative routes have a higher hop count, the FR network is used.

If PVCs are being used, the connection is always available. If SVCs are used, the definition of each node along the FR network specifies the FR address to obtain a connection. The SVC connection is established if a request demands use of the connection. This is automatically established by the APPN node at the access point.

IP nodes can use one of several routing algorithms to provide the necessary route determination. Although dynamic routing can be used, normally predefined route entries would be created for the dynamic connection.

The normal predefined route process would be used. The fact that the link between Router A and Router B in Fig. 10.2 is across an FR network does not have any impact. The routers only see that a connection exists and that packets can be passed between them.

In this case, normal route discovery and determination is used to calculate the route between Workstation A and Workstation B. The determination of which route is chosen is partially dependent on the routing algorithm that is being used and the definitions of the connections. In the case shown in Fig. 10.2, if RIP is being used, the FR network is the optimal route because the hop count is lower than the route through the routers that are ethernet connected. One of the alternative routing algorithms may make a different determination.

10.1.2 Virtual point-to-point connection

The FR network creates the image of a series of point-to-point links. Every destination is viewed as being one hop away. This is similar to the concept of a connection network as provided in the APPN architecture.

Because the nodes are only one hop away, there is no requirement to provide routing between the local entry point to the FR network and the remote exit point. The only routing that is needed is for the

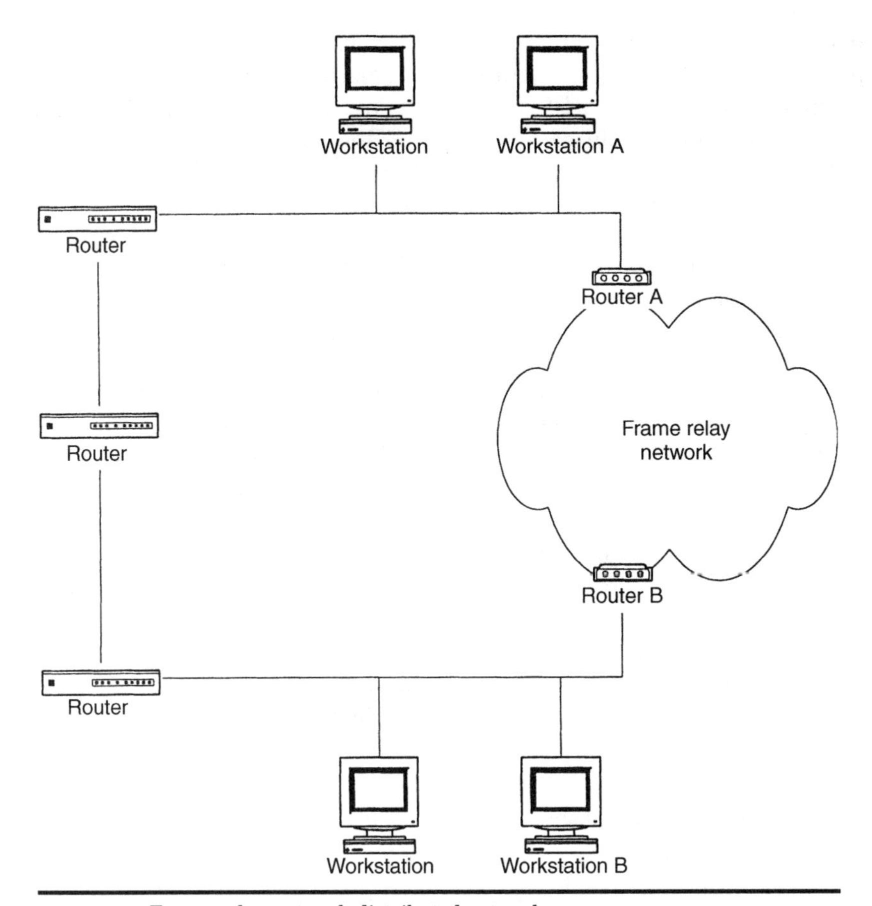

Figure 10.2 Frame relay network distributed network.

purpose of determining the entry point to the FR network. If multiple interfaces exist, each of these is viewed as being directly adjacent to the destination. Thus, the hop count is immaterial in the route determination.[2]

10.1.3 Bandwidth on demand

Unlike most other serial communication methods, FR networks provide for the ability to exceed the subscribed bandwidth for a short

[2]The speed of the respective interface, the distance to the access point, and other defined variables (such as security) are the parameters for route determination.

period of time. These *bursts* of traffic are characteristic of the transportation of certain types of traffic, such as that from LANs.

On subscription for a FR PVC or during call setup for an SVC, an committed information rate (CIR) is defined. This CIR is the maximum sustained data rate the network provider guarantees information can be supported. If the actual flow to the FR network is below this figure, no congestion should result.

The connection parameters include the specification of a burst rate. This is the maximum amount of additional traffic, over and above the CIR, the subscriber will transmit into the network for a given period of time.[3] The purpose of this specification is to set a maximum transmission level for the subscriber. If data is sent into the network at a higher level, the network is free to discard all of the excess traffic.

The variation between network providers is even larger than data in excess of the CIR when discussing this level. As is true of a rate in excess of the CIR, some network providers have been attempting to support the additional data rate. In this case, the network provider passes the data as long as there is no impact to other subscribers and the network has the necessary capacity. Other providers stop accepting data when the burst rate is exceeded.[4] In this latter case, the rates are strictly enforced. This ensures that the provider can properly configure its network and results in a more businesslike manner of operation.

10.1.4 Discarding of messages

If the amount of traffic exceeds the CIR, the FR network provider attempts to deliver the traffic successfully. If the network provider can support the excess traffic, most providers attempt to successfully deliver the traffic. If the network cannot support the extra traffic, it initiates a selective discard of messages.

The first messages that are discarded are frames that have the *discard eligibility bit* (DE) set (DE = 1). These frames are designated as being traffic that can be discarded.

When the level of excess data rate exceeds the maximum burst rate, it is also possible for nonmarked traffic to be discarded within the FR network.[5]

---

[3]The exact period of time varies by the FR network provider.

[4]This second option is quickly becoming the rule as FR has become popular and network capacity is being achieved.

[5]My experience shows that some FR providers enforce the CIR and burst rate by discarding all traffic that exceeds the burst rate.

10.1.5 End point error recovery

Frame relay networks do not provide for link-layer transmission recovery. This is a result of the reduction in the error rates for communication interfaces. Since these rates have come down, the benefit derived from performing error recovery within the network is reduced. The tradeoff for having error recovery performed by the endpoints must be weighed against the cost of transmitted data across connections that may have already successfully transported the data. In the case of FR, a decision has been made that the amount of retransmissions is sufficiently low that the cost of supporting additional data transmissions is tolerated.[6]

Frame relay moves the point of error recovery detection and retransmission to a point outside of the FR network. As a result, the links are able to operate without the overhead associated with providing error recovery. Figure 10.3 shows a comparison of link-level and network-level error recovery.

Movement of the error detection and recovery to outside of the FR network, requires that a message be retransmitted through the entire network. Though this has a higher overhead for the packets that need error recovery, the frequency is so low that there is little or no impact to overall throughput. In addition, there is often a much higher user data rate for FR as compared to the leased line network, which often is the target of conversion.

10.1.6 Congestion notification

Frame relay networks can become congested from multiple locations exceeding their CIR. When congestion occurs, the FR switch has two methods of notifying nodes of the difficulty. These methods use the backward explicit congestion notification (BECN) and forward explicit congestion notification (FECN) bits within the FR header.

The BECN notifies downstream devices of the congestion within the FR network. These devices should then slow the introduction of data into the FR network. By reducing the transmission of data into the network, the congestion should be reduced or eliminated.

The FECN notifies upstream devices of congestion within the FR network. Although the upstream device may not be the direct cause of the problem, it can use higher-layer protocols to signal the down-

[6]Users see the cost of the additional data transmissions as overweighed by the benefit of the customary increase in data rates exhibited by most FR networks. If a 9600-bps leased connection is exchanged for a 56K bps FR connection and if some data has to be retransmitted, the user still has a much higher throughput.

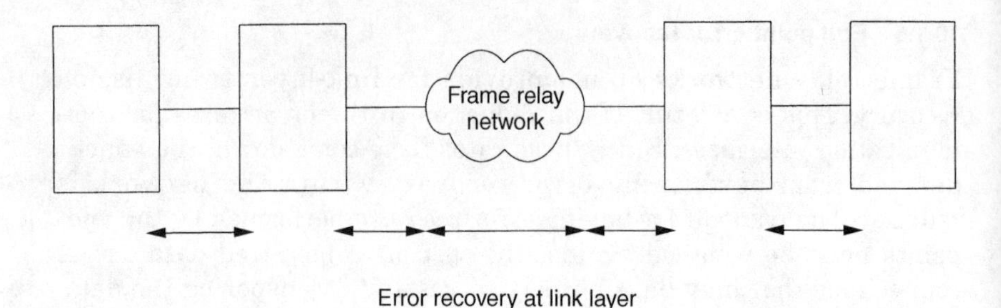

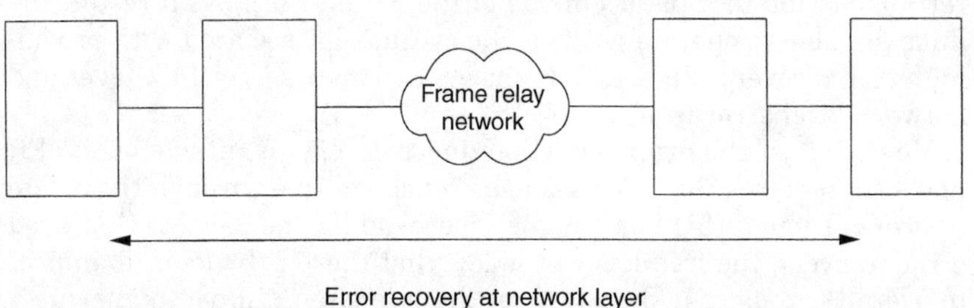

Figure 10.3 Mixed topology network.

stream device to slow data transmission or the upstream device can respond slowly to receiving data, which can reduce the data throughput and alleviate the problem.

The DE bit in the FR header allows for the identification of frames that can be discarded by the FR network in the case of congestion. Unfortunately, there is no "standard" way of interpreting this bit.

In some networks, the DE bit is set by the user as frames enter the FR network. The FR network uses this bit to determine whether the frame is eligible for discarding.[7] In this case, the network recognizes the setting made by the user to differentiate frames.

In other cases, the FR network operates on frames in an ad hoc manner. In this case, the network discards frames not on the setting of the DE bit, but on the conditions at the time of frame arrival. If congestion is indicated at the time of a frame's arrival, it is discarded. This method does not take into account the designation made by the user.

---

[7]All networks override this bit if the congestion is sufficient to warrant widespread frame elimination.

A third method works in conjunction with the CIR. If the input rate is below the CIR, the network does not modify the DE bit. On the other hand, if the rate exceeds the CIR, the network overrides the user specification and sets the DE bit on all frames that exceed the CIR. This method operates so that the CIR-rate is respected, along with the user designation of the DE. Once the rate exceeds the CIR, the network has the option of discarding frames.

10.2 Introduction to RFC 1490

As has been stated, RFC 1490 defines an encapsulation scheme for various protocols that can be transported across a frame relay network. This RFC does not describe any type of protocol conversion nor does it define a method of providing equivalent functionality in a transport other than frame relay.[8]

RFC 1490 creates a transparent transport that can be used by an assortment of protocols for connectivity. As was stated in Chap. 8, this type of encapsulation scheme presents some problems when the transport includes some link-level messages. Although RFC 1490 makes no special provisions to overcome the problems associated with an encapsulated interface, implementers of this RFC keep these link-level messages off the encapsulated interface. Thus, the implementers of this RFC have implemented spoofing of link-level messages outside of the FR network.

10.2.1 Routed frames

RFC 1490 defines two types of message formats. The first is for routed frames. These are frames that are output from a router that implements RFC 1490. These frames are in one of two forms which are shown in Figs. 10.4 and 10.5. The distinction between these two forms is that if an NLPID is assigned, there is no need to include the following 6 bytes:

- Pad field (1 byte)
- SNAP header (5 bytes)

In the case where an NLPID is not assigned, a NLPID of 0x80 is used. This signifies that a SNAP (subnetwork access protocol) header is used. An OUI of 0x00 00 00 is specified in this header, along with the Ethertype designator, followed by the actual protocol data.

---

[8]This is the case, though it would not be difficult to extend this RFC to support equivalent transports, such as X.25.

Q.922 address
Control x′03′
Pad
NLPID
OUI
PID value
MAC
User data frame
LAN FCS (if PID specifies its use)
FCS

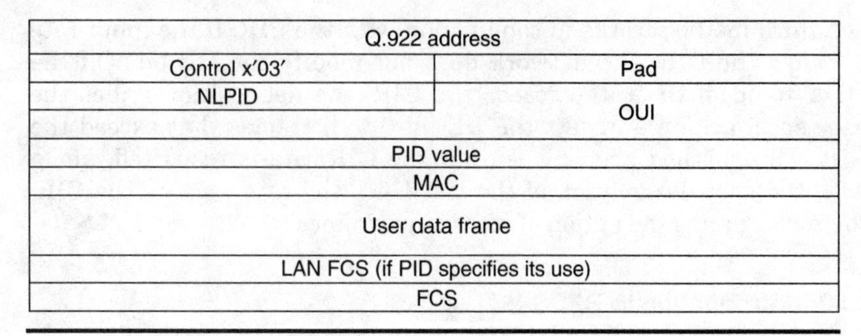

Figure 10.4 Comparison of error recovery procedures.

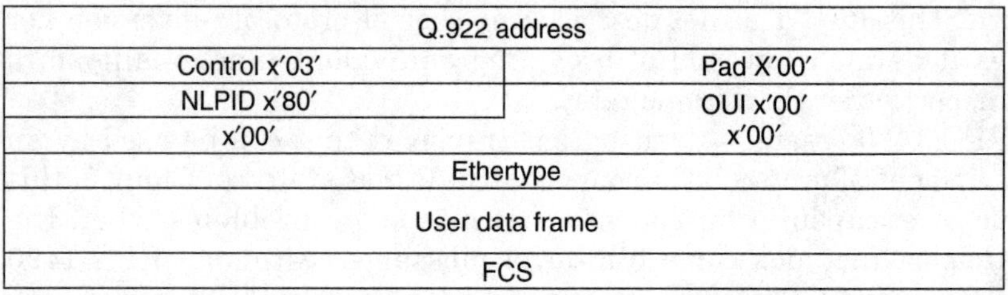

Q.922 address
Control x′03′
Pad X′00′
NLPID x′80′
OUI x′00′
x′00′
x′00′
Ethertype
User data frame
FCS

Figure 10.5 Format of RFC 1490 frame.

10.2.2 Bridged frames

The FR network can also be used as a medium for bridging data traffic between LAN segments. In this case, the data traversing the FR network is output from a LAN bridge that allows frames to be passed between LAN segments. In this case, the FR network becomes a virtual LAN segment along the path.

Figure 10.6 shows the base frame format. Note that the PID used in the frame is obtained from the list in Table 10.1. Thus, this field is

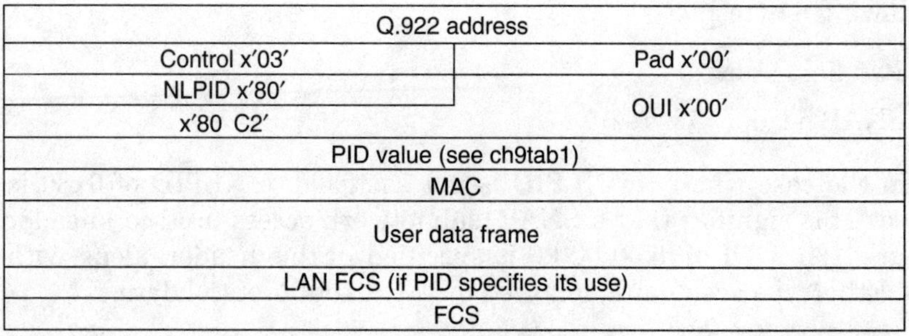

Q.922 address
Control x′03′
Pad x′00′
NLPID x′80′
OUI x′00′
x′80 C2′
PID value (see ch9tab1)
MAC
User data frame
LAN FCS (if PID specifies its use)
FCS

Figure 10.6 Format of RFC 1490 routed frame.

TABLE 10.1 PID Values for Bridged Frames for RFC 1490

PID Values for OUI x'0080C2'		
With FCS	**Without FCS**	**Media**
x'0001'	x'0007'	802.3/Ethernet
x'0002'	x'0008'	802.4
x'0003'	x'0009'	802.5
x'0004'	x'000A'	FDDI
	x'000B'	802.6

0x00 01 for Ethernet frames with the FCS, and 0x0007 if the FCS is excluded from the frame.

10.3 How Does RFC 1490 Work?

An FR network that utilizes RFC 1490 allows the creation of a transparent network that enables the transportation of an assortment of frame formats and protocols. Each of the datastreams is segregated on a different VC. Through the use of the encapsulation header, the receiving station is able to determine not only the routing of the frames, but also enables the frame to be decoded properly.

The result of this encapsulation/decoding is the creation of a virtual network that supports the transportation of almost any communication frame. The link-level transportation facility can form the basis of a shared common media for the transportation of data. Often, this type of a network forms the basis of a common backbone network.

At the periphery of the network is an assortment of equipment that provides either the bridging/routing of 802.2 or Type II frames or equipment that can convert user data into one of these forms, so that it can be supported for transportation. Figure 10.7 illustrates a typical network that uses this design.

The conversion of frames into this standard format can provide additional services. These services can include spoofing of some of the noncommon protocol's link-level messages, as was described in Chap. 8. The added services can also include network management extensions or the implementation of a private protocol to implement extended facilities.[9]

[9]Some vendors provide the ability to multiplex traffic from several interfaces. Through the use of a private header that is prepended to the datastream, the data can be demultiplexed at the remote end and transmitted to the correct destinations.

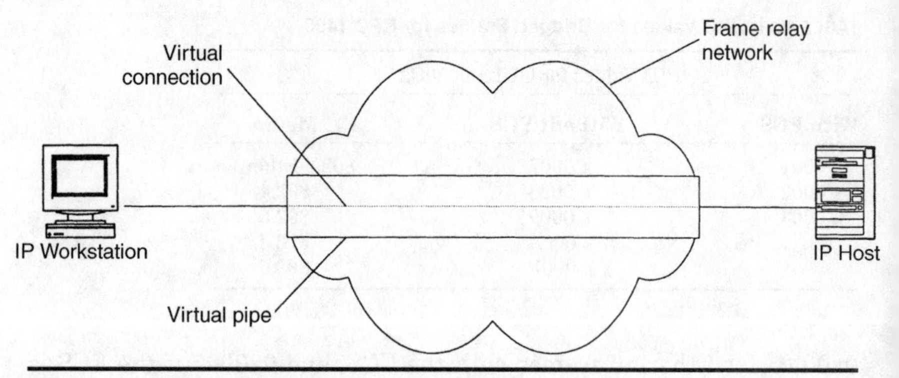

Figure 10.7 Creation of virtual connection across FR network.

10.3.1 Fragmentation Through the Frame Relay Network

Frame relay networks, generally, have a much smaller frame size than most of the 802.2 and Type II LAN networks. As a result, a method of fragmenting transparent messages across the FR network had to be devised. The design objective was to make the fragmentation only applicable to the FR network. In this way, the protocols passing through the FR network would not have to be reconfigured nor would they even have to be aware of the transport that was being used.

The result is a method of fragmentation that is limited to the periphery of the FR network. Only these nodes are involved in the fragmentation and reassembly of data.

The fragmentation procedure is a two-stage method that allows any of the RFC 1490 frames to be fragmented. In the first stage, the frame is handled normally by determining the frame type and applying the RFC 1490 header, as was previously described.

In the second stage, a determination of fragmentation is made. If the created frame is larger than the MTU of the FR network, the frame must be fragmented.

To accomplish this, another header is applied to the frame. This is the RFC 1490 fragmentation header. It is very similar to the bridged header, except that the OUI is set to '0080C2' and the PID is set to '000D.' An addition of 8 bytes are used within this header to provide a sequence number, offset, and a fragmentation complete bit.

The sequence number is a two-byte field that is initialized to a random value and incremented by 1 for each segment of a message. The offset assists in the reassembly of the original message. It is an 11-bit field that is initialized to zero and which contains the byte offset of the fragment divided by 32. For this reason, each frame transmitted through the FR network should contain an even number of 32 byte segments. The final bit is set only on the last frame of a segmented message.

If the receiver detects any problem during the reassembly of a message, the reassembly action is terminated and the partially assembled frame and additional message segments are discarded. Because the RFC 1490 entity is not involved in the higher-level protocol, there is no backward indication that the message has been discarded. It is the responsibility of the end station to determine if the message has arrived and to initiate a request for retransmission, if necessary.[10]

Figure 10.8 shows a large message before fragmentation. Figure 10.9 shows an IP frame that has been fragmented in this manner. Note that the final bit is only set in the last frame of the segment and that the offset value is initialized to zero, in the first frame. In the second (and last) frame, the offset value is divided by 32.

The result is that the IP entity is unaware of the fact that the frame has been segmented. Both of the endpoint IP devices are unaware of the manipulation of the frame by the RFC 1490-compliant resources at the periphery of the FR network. In addition, the endpoint IP devices are also unaware that the frame had even been transported across a FR network. They are given the view that a normal 802.2 or Type II-compliant networks have been utilized throughout the span between the IP devices. This view can be seen in Fig. 10.7.

[10]It is this lack of error detection that provides the smaller and faster processing provided by an FR network. The counterbalancing side is that error detection and recovery is provided by the periphery of the network. If this type of activity is infrequent, the impact on the FR network is minimal.

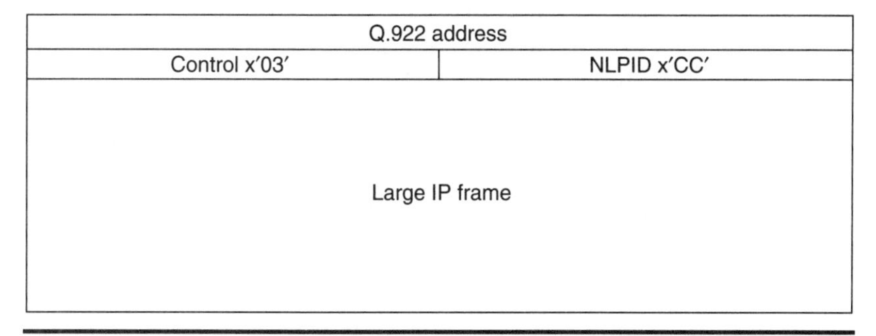

Figure 10.8 Large IP frame *before* fragmentation.

Q.922 address
Control x′03′ | Pad x′00′
NLPID x′80′ | OUI x′00′
x′80′ | x′C2′
PID x′00 0D′
0 | RSVD | Offset from start (0)
Imbedded CTRL x′03′ | NLPID x′CC′
First *m* bytes of user data frame
FCS

First segment of IP frame

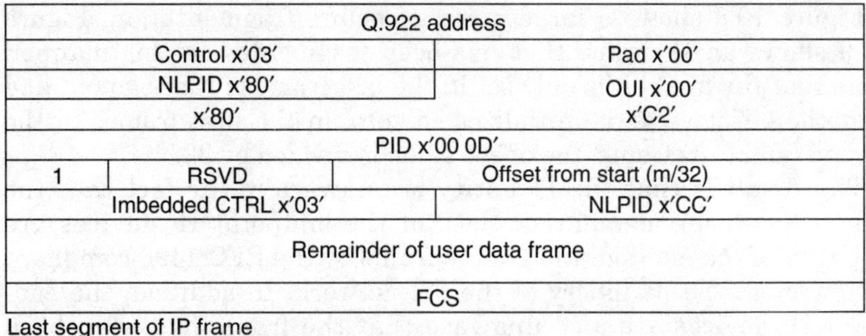

Figure 10.9 Segmentation of IP frame using RFC 1490.

10.4 Integration Using RFC 1490

To use RFC 1490 for the integration of different protocols, you must first determine if a FR network is available and viable to meet your networking requirements. The determination can be made by ascertaining whether FR is even available at the sites that you need connections. Assuming that FR is available, you must also determine whether conversion to an FR network would gain you competitive or budgetary advantages. If the conversion will result in higher costs or has other adverse effects, it obviously is not advisable to move forward with a change in network connectivity.

You must also determine if you can pass the desired datastreams through an RFC 1490-compliant product. If there is no standard conversion to a required frame, it will be chancy if a standard interface can be used. Instead, you may require a customized interface to be built that can meet your networking requirements.[11] This can result

[11]It is possible that all of your interfaces pass this test except for one. In this case, it may be necessary to build a customized interface for only that single exception; the rest of the network would adhere to a standard implementation.

in some limitation on what products can be utilized for that particular connection and where they can be utilized, but it should cause no similar limitation on any other interface.

When the connection for the nonstandard interface is started, it utilizes a particular DLCI. This connection is paired with a similar piece of equipment on the other side of the FR network that understands how the data has been encapsulated. This paired component is able to extract the original data from the received frame and transmit it to the destination.

Standard encapsulation is still fully supported with no modification to any other component on either side of the FR network. Data is received, the header understood, and the normal extraction proceeds, along with the data transmission to the actual destination. These standard interfaces see no differential from their normal processing. This is because the nonstandard datastream is isolated to specific circuit(s). All other circuits have no awareness of what is carried on the other circuits, so there is no effect.

10.4.1 Address resolution

The DLCI number has only local significance. There is no global numbering scheme that you can expect to be carried out throughout a specific FR network. This results in an inability to utilize the DLCI as making an indication of what type of encapsulation is provided or the partner of the connection. Yet there are many cases where this information is needed. As a result, methods of supporting address resolution were developed in RFC 1490.

Though a DLCI has only local significance, it will stay consistent throughout a connection. Thus, if DLCI = 40 is used on one side of the connection and the other side used DLCI = 50, these numbers will stay consistent throughout a connection. Utilizing this fact, and equating this DLCI to a hardware address, it is possible to utilize an ARP procedure to provide and complete the address resolution.

Figure 10.10 shows how the fields within an ARP request and response can be used for this purpose. By using the DLCI as the source hardware address, it is possible to fill in the respective source address fields and enable the partner to likewise determine the pairing of DLCI between them. This provides a resolution of "hardware address" to "protocol address." The reverse address resolution protocol (RARP) is also supported in a similar fashion.

10.4.2 Transportation of TCP/IP

IP data is encapsulated for transport through an FR network by using the NLPID of 0xCC, rather than the routed form of encapsula-

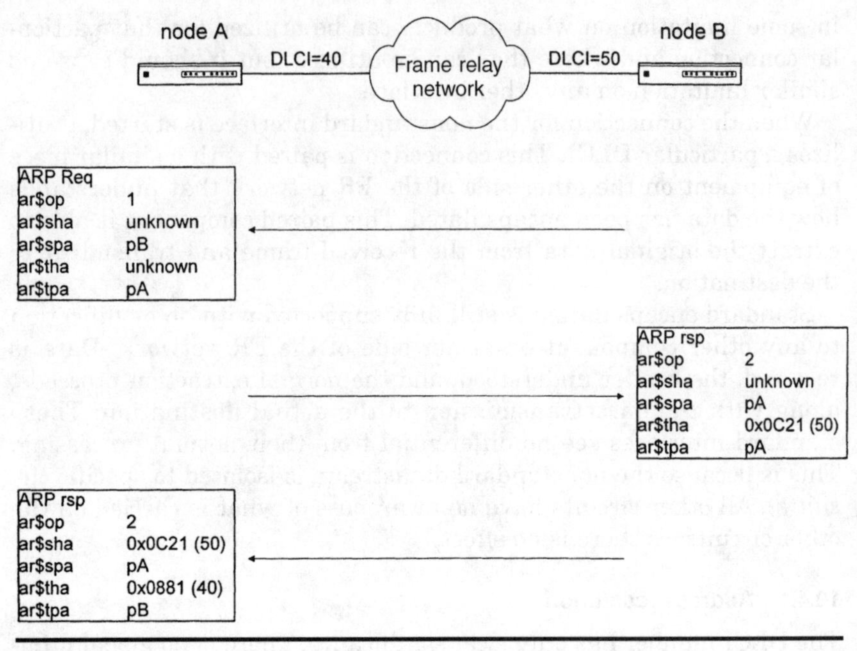

Figure 10.10

tion using an NLPID of 0x80 and a SNAP header. By using the NLPID-0xCC form, 6 bytes of header data are removed. This reduced overhead allows for a more efficient transport of data plus the use of a unique NLPID reduces the overhead on decoding the frame.

As this form of encapsulation is consistent with the form used for X.25 (RFC 1356), this eases the implementation and development required to provide support by making RFC 1490 similar to the encapsulation provided in an earlier form.

In this form, the total overhead (over a unencapsulated frame) is 4 bytes. This is a very low level of overhead to be passed across the FR network.[12]

[12]Chapter 11 will describe the overhead associated with other types of encapsulation and bridging. You will see that the other forms, without exception, require more overhead for the encapsulation header.

10.4.3 Transportation of SNA

Subarea and APPN SNA data is transported across an FR network by using extensions to RFC 1490.[13] These extensions describe how SNA data is encapsulated and passed across an FR network.

This method uses the CCITT Q.933 NLPID of 0x08. The four bytes following the NLPID define the layer 2 and layer 3 protocols being used. Table 10.2 shows the fields used for layer 3 encoding.

The layer 2 encoding is provided by specifying 0x4C 80.[14] This identifies the frame as conforming to 802.2.

The layer 3 encoding defines the type of data that is encapsulated within the frame. Because there is an identification of FID4, peripheral FID2, and APPN FID2, it is possible without even looking inside the frame to determine what type of data is contained.

10.5 Limitations of RFC 1490 Implementation

It would appear that RFC 1490 is an answer to the problem of a multiprotocol network. By providing the foundation for the transportation of multiple protocols while requiring a minimum amount of additional overhead, the RFC 1490 solution seems to be a solid answer to this problem.

As with most things in life, things are not as good as they first appear. In this case, there are several limitations on the delivery of this efficient methodology. Although limitations exist, they are not

[13]FRF.3, *Multiprotocol Encapsulation Implementation Agreements.*

[14]The trailing 0x80 is a pad field and has no protocol significance.

TABLE 10.2 Codepoints for Encoding of SNA Data Types

Code Point	Description
0x81	FID4 SNA
0x82	Peripheral FID2
0x83	APPN FID2
0x84	NETBIOS

onerous; there are just some limitations on the environment in which you can utilize this interface.

10.5.1 Transportation limitation—FR only

The first limitation is that RFC 1490 is only available in conjunction with an FR network. There is no current provision for transporting this interface to other link protocol environments.

Although FR is becoming available to more locations, this does not mean that it is universal. As a result, if FR is not available, RFC 1490 cannot (by definition) be utilized.

10.5.2 Alternate access port to FR network

Although FR networks provide alternate routing through the FR network, alternate access to that network is not provided. This can be provided in one of several paths, which include:

- Dial access port
- Secondary leased connection
- SVC through FR network
- Integrated Services Digital Network (ISDN)

The first method uses a dial port to an access point into the FR network. This dial port has a leased line access to the FR network, but a dial interface to the access node itself. With this dial backup node, the FR network could still be accessed, but a routing determination would have to be made in how to access the dial port. See Fig. 10.11.

This method is useful if the local access to the FR network is broken. A local node makes the determination that the local access point is unavailable and that dial access is desired. This node, or a node associated with this node, makes a dial connection to the defined backup node. This node has a leased connection to the FR network. A routing decision is mandatory to provide a path to the backup access node, but the connection to the remote destination would proceed as it does normally. New routes would be necessary for all ports that needed FR access, but this could be provided by the network-layer routing protocol being used.

The second method uses a second access port to the FR network. Again, a routing decision would be necessary to determine the route

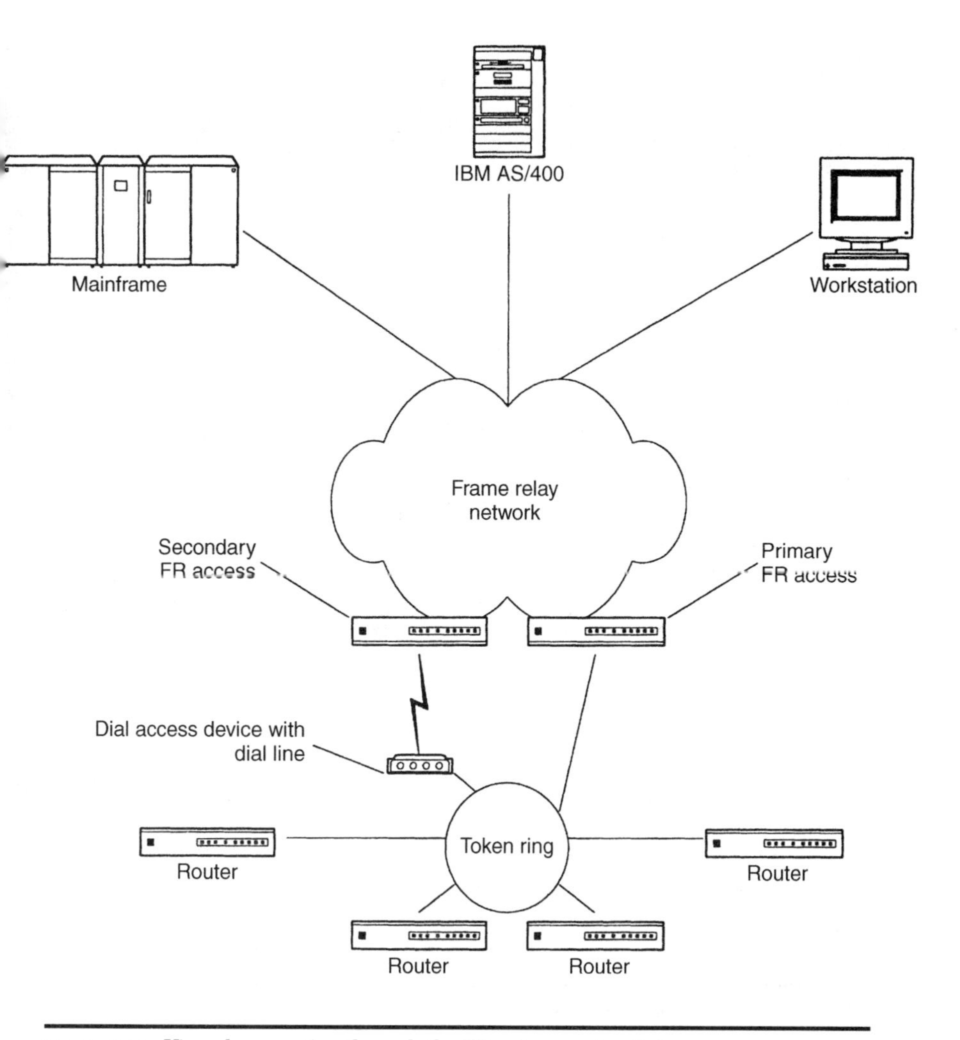

Figure 10.11 Virtual connection through the FR network with dial access.

to the alternative path into the FR network. This method is also used when the local access point to the FR network fails. The difference is that in this case, access to the access point is along a leased facility. See Fig. 10.12.

This solution allows for the use of a secondary access point to the FR network. This alternate access point can be utilized in the normal flow as a load balancing point to the FR network. When the primary

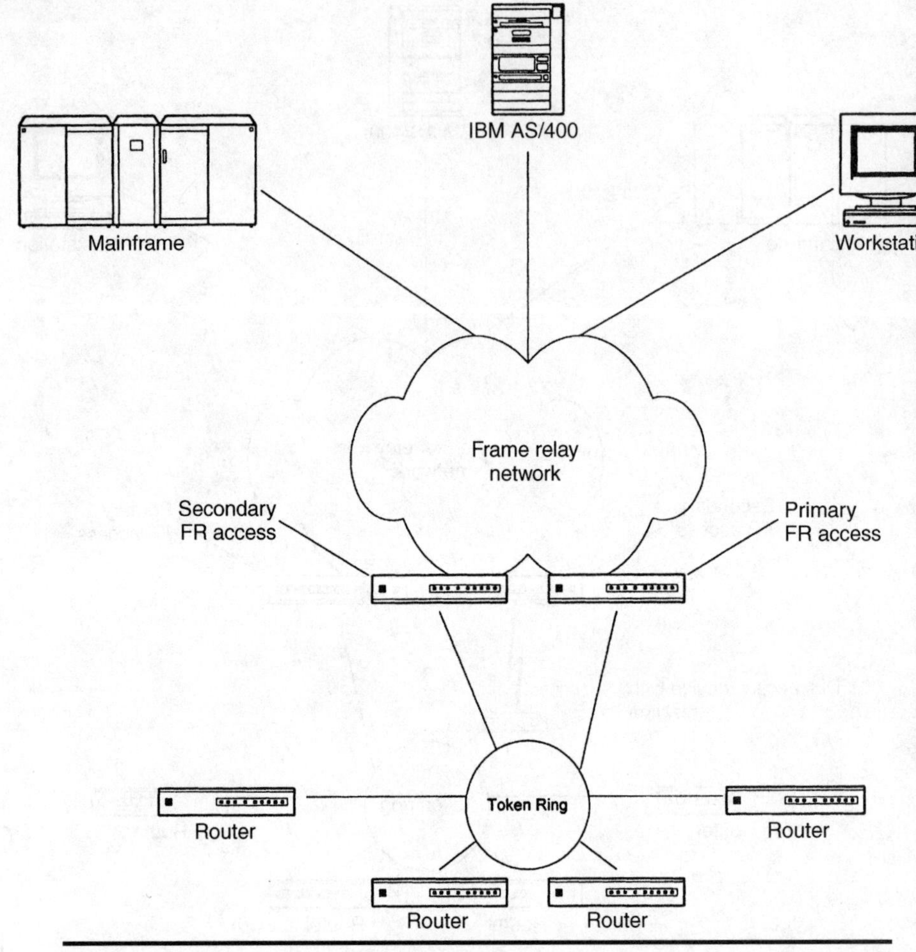

Figure 10.12 Virtual connection through the FR network with leased line.

path to the FR network becomes unavailable, this path can take over the load until the primary can be reestablished.

A third method uses SVC connections[15] to the FR network. These SVC connection can be dynamically established to provide connectivity. The creation of these SVC connections can be triggered by the

[15]This is more of a theoretical solution as few FR providers currently support SVC operation.

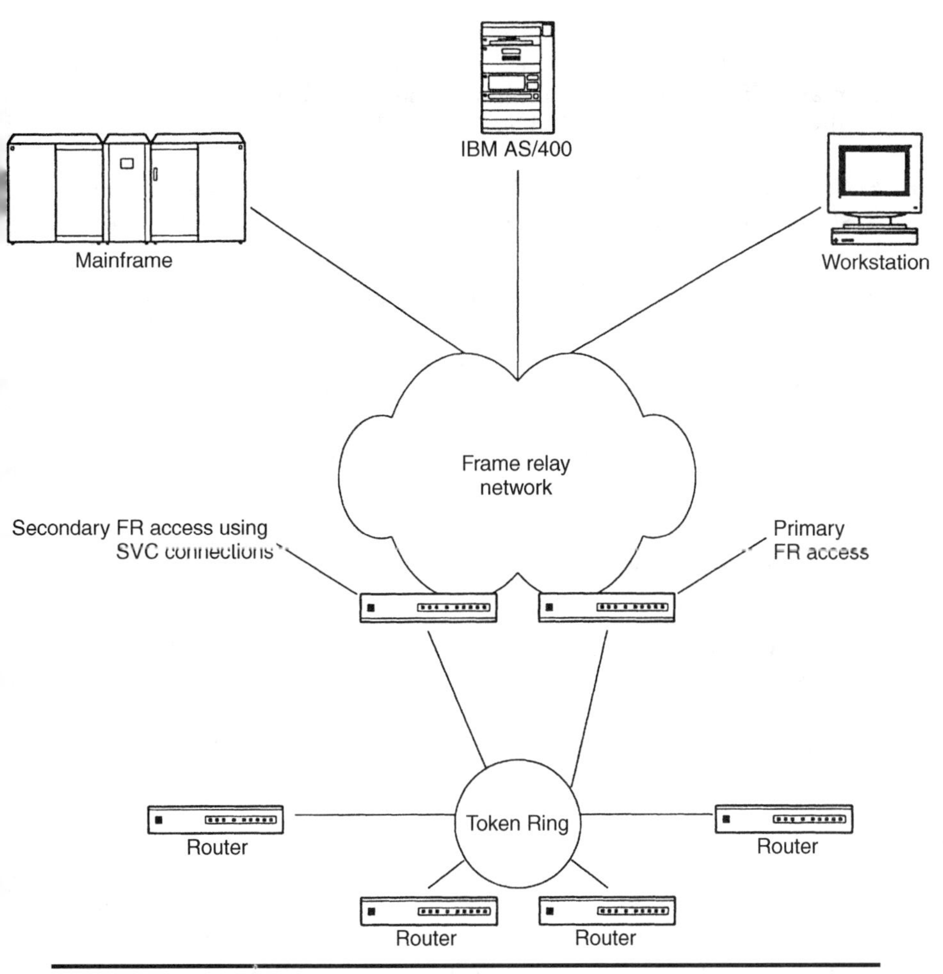

Figure 10.13 SVC connection to FR network.

loss of normal PVC connections on the primary access node. See Fig. 10.13.

By using SVC rather than PVC, this solution provides the same connectivity as in the previous method, but can reduce connection costs by not requiring the connection to be permanently available. On the other hand, this solution would require SVC support in both the FR network and the access device. As there are few implementations that support SVC usage, this solution may be difficult to configure.

Another method uses basic-rate[16] ISDN access to the FR network. This is becoming particularly important since ISDN is becoming increasingly available. The interest is because of many factors associated with ISDN. These include:

- *Switched connectivity.* ISDN access uses a switch access that is similar to normal analog dial access. As a result, costs are incurred only while the switched ISDN connection is active.[17]
- *Supported speeds.* ISDN basic-rate access allows much higher line speeds than analog dial access. Even using the currently highest speed analog technology, V.34, you are limited to a line speed of 28.8K bps.[18] Basic-rate ISDN allows for two 64K bps connections. Although these connections can be combined into a single 128K bps circuit, even one 64K bps connection provides for far higher throughput than using V.34 on an analog circuit.[19]
- *Digital.* ISDN uses digital technology for connectivity. As a result, much higher reliability than normal analog dial access is achieved.

10.6 Network Management Through RFC 1490

Management of the FR interface consists of two parts. These are:

- The logical connection between the end units
- The FR network itself

The first piece is provided by the same management interface as is currently utilized. From the logical connection point of view, the existence of the FR network for connectivity is not interesting. The network should be transparent to the existence of the FR network in the middle.

---

[16]Basic-rate ISDN access uses a 2B + D interface. That is, two 64K bps lines, plus a 16K bps line that is normally utilized for control information. This contrasts with primary-rate ISDN, which provides for a 24B + D or 30B + D connection.

[17]There are setup and monthly service charges, but I am overlooking this portion of the cost in my analysis.

[18]Compression technologies allows greater throughput than this, but the connection operates at this speed.

[19]Similar compression techniques can be utilized to provide the same advantages as with the V.42 compression supported by V.34.

The second part is provided by the local management interface (LMI). This is an interface to the FR network that allows information at the subscriber network interface (SNI) to communicate, bidirectionally, as to the status of the partner. This is a component of the consolidated link-layer management (CLLM) that provides status information to be conveyed to and from the FR network. Through this interface, both the network and the CPE can determine the status of the interface.

DLCI 1023 is reserved for use by the LMI. Information about the DLCIs used between the CPE and network is communicated within the LMI message. This message is sent as a HDLC UI (unnumbered information) frame.

The information that can be ascertained can enable a management platform to gain a great deal of information about the status of the FR network. This includes detailed information on the network's view of why an error or failure occurred. Some of the values of the diagnostic field are shown in Table 10.3.

Through the use of the LMI interface, the local interface can gain information concerning the status of the VCs within the FR network. With this information, the logical protocol network can be superimposed to create a consolidated view of the network that includes the end devices, the status of the logical session between them, and the network upon which the session is riding. This provides a very rich environment for network management, if the management platform is capable of utilizing all of the information and provide a consolidated view.

TABLE 10.3 Cause Codes for Frame Relay LMI

Bits								
8	7	6	5	4	3	2	1	Cause Description
0	0	0	0	0	0	1	0	Network congestion, short term
0	0	0	0	0	0	1	1	Network congestion, long term
0	0	0	0	0	1	1	0	Facility or equipment failure, short term
0	0	0	0	0	1	1	1	Facility or equipment failure, long term
0	0	0	0	1	0	1	0	Maintenance action, short term
0	0	0	0	1	0	1	1	Maintenance action, long term
0	0	0	1	0	0	0	0	Unknown, short term
0	0	0	1	0	0	0	1	Unknown, long term

10.7 Summary

Frame relay networks provide a new methodology of creating a backbone topology. It has several advantages over existing leased line facilities. The first of these is the creation of virtual backbone topology. This results from the ability of the FR network to create a flat, single-hop topology between nodes connected to the FR network. This is similar to the topology resultant from dial, X.25, and some LAN configurations.

The advantage frame relay has over the other serial alternative is its variable data rate. Frame relay supports two levels of data rate. The base configuration is the committed information rate (CIR). This is the data rate at which the FR network provider guarantees throughput. The second rate is a burst rate. This rate is supported to the point that the network has available capacity. The purpose of this rate is to support short periods of data throughput requirements above the CIR. These spikes in data rate result from the nature of networks and the users of those networks. In many cases, users of the network have a normal level of requirement, but also have peaks. These peaks could result from such activities as file transfers or just a period of high need.

Although the FR network guarantees to be able to support data at the CIR rate, if FR customers exceed this rate, it is possible that the network will be unable to deliver messages. In this case, FR enables the network to discard excess traffic.

FR networks do not provide for the recovery of frames. Responsibility for error recovery is moved outside of the FR network to either the end stations that are using the FR network or access devices attached to the network. It is the responsibility of these devices to recognize a data error and to provide a method of correcting the error.[20]

RFC 1490 is based on an FR topology. The RFC defines a header format that allows for the transparent transportation of multiple protocols.

The headers fall into two categories. The first is a *bridged format.* In this case, the frame contains DLC information as well as protocol and user data. The second format is a *routed format.* This is used when the frame has or will be routed through the FR network. This

[20]Often this is within the domain of a higher level protocol, but some cannot readily fulfill this requirement. In these cases, recovery is provided by the device and isolates the knowledge from higher-level protocols.

frame eliminates the DLC information from the frame that is transmitted through the FR network.

RFC 1490 defines how several address resolution protocols can be supported.[21] These include ARP and RARP. The method is based on the fact that, although DLCIs are not consistent across an FR network, a connection uses a consistent set of DLCIs. By viewing the DLCI number as a hardware address, it is possible to provide a mapping between this "hardware" address and the logical address associated with the resource. Thus, although a LAN device does not have a DLCI, in the view of the remote device this correlation exists.

Internet Protocol data is encapsulated for transport through a FR network by using the NLPID of 0xCC in a routed frame format. This makes the FR encapsulation format for IP consistent with other transport topologies, such as X.25 in RFC 1356. This allows the use of a common code point for both network topologies and speeds the availability of the IP encapsulation for RFC 1490. It also reduces the code required to support these different topologies.

SNA encapsulation across FR has been defined by IBM using a Q.933 message format. This uses an NLPID of 0x08. It also defines the use of a 4-byte header to provide definition of the layer 2 and 3 protocols. Layer 2 is consistently encoded as '4C80'. This defines the frame as adhering to 802.2 format. The layer 3 protocols are defined by four code points. These identify the type of data that is encapsulated and allow for quicker determination and routing of frames. The types are:

- FID4 INN traffic
- FID2 peripheral node—PU2.0
- FID2 peripheral node—PU2.1
- NetBios

Although RFC 1490 provides many advantages, there are also limitations to its implementation. The most significant limitation is that this RFC only defines usage for FR networks. If FR is not a viable alternative as a network topology, RFC 1490 connectivity cannot be used.

Network management of the FR interface is provided at two levels. The first is isolated from the use of frame relay. At this level, the FR

[21]The protocol is not unique to RFC 1490, only the method of implementation.

network appears only as a connection point between resources. At this level, an existing network management interface can be used.

The second level allows for information concerning the FR network to be interrogated and interpreted. This level uses the LMI to the FR network. This interface allows information at the SNI about resources within the network and VCs through the network. This allows for bidirectional information flow. It is through this interface that information such as network congestion is provided.

Chapter

11

Data Link Switching (DLSw)

This chapter briefly reviews the design for data link switching (DLSw) and analyzes the effects of some of the design points. This chapter will reiterate all of information found within the DLSw RFCs, RFC 1434, and RFC 1795. It is meant to be a tutorial, not a thesis on the design.

11.1 Background

More and more enterprises are using IP as their backbone network architecture. This situation has resulted, especially within the United States, because of the low cost of connection devices and the increased availability of software that supports the IP architecture.

Another contributing factor has been the explosion of client/server applications. Many of these applications have been developed on UNIX platforms that provide IP as their native network architecture. These client/server applications are often "isolated" to a specific department within a larger enterprise. This isolation allows the department to install a complete solution that includes not only the applications, but also the network infrastructure, including leased circuits and routers to provide connectivity.

As the application is installed, requirements arise for connectivity to other applications and databases. Because connectivity is being provided by an assortment of departments and individuals, hardware and software from a variety of vendors is often included. Though a solution may work in isolation, problems arise when integration of the different vendor solutions is attempted.

11.2 Objective of DLSw

This is the background that led to the creation of DLSw. The heterogeneous networks that companies have built require that a method of integration be agreed to between communication equipment vendors. Of particular interest are vendors that provide SNA connectivity.

The reason for this interest is multifaceted. The first reason is the size of the IBM connectivity market. Although there are more IP devices in use, the IBM market represents the majority of large commercial customers.[1] To put it simply, this is seen as a large money-generating marketplace.

Each of the vendors in this market has created its own method of providing SNA connectivity. Many of them operate quite affectivity and efficiently. Unfortunately, these methods do not work together! This has resulted in many connectivity problems for companies after they have invested in an assortment of network vendors' equipment. Though these vendors have a certain vested interest in keeping their customers closely tied, they still do not want their customers upset or concerned that they are being sold a totally closed system that the customer will not be able to operate with. As a result, DLSw was created to provide a standardized method of providing this connectivity.

11.2.1 Common method for transport of SNA over TCP/IP

The achievement of a common method of supporting the transport of SNA data across on IP network was the driving force for the creation of DLSw. This process had to allow for participation of all vendors that were interested.

To this end, IBM provided an informational RFC describing a method of transparent SNA transport across an arbitrarily sized IP network. The IETF incorporated this information as RFC 1434 in March 1993.

This RFC described how IBM had itself provided for this transport. IBM made the technique available to the public in an effort to get it accepted as the "standard" method for SNA transport across an IP network. IBM also created a related interest group (RIG) within the APPN Implementors Workshop (AIW) in a further effort to standardize a methodology for SNA transport.

IBM's efforts paid off within the AIW RIG, in that the group took up the charge and further designed the transport. The participants in

---

[1] At least, a majority of these customer have this requirement.

the RIG recognized that the IBM design had areas that needed improvement[2] and proceeded to provide for those enhancements.

11.2.2 Implementable

The design provided in RFC 1434 did not rely on any specific proprietary interface for the SNA transport. This agreed with one of the driving design points that specified that only standard interfaces were to be used so that the design was as open as possible. Though new messages were implemented between devices that implemented DLSw, there were no private interfaces utilized. Everything within the design was available to any vendor that wished to participate in this transport.

11.3 Overview of DLSw

SNA and Netbios are session-oriented protocols. As such, they use 802.2 LLC2 protocol on a LAN. LLC2 uses fixed timers to detect the loss of frames. Because these timers are LAN-oriented, they assume that connectivity speeds are reasonable for LAN connectivity. If a WAN is placed between the session partners, these timers often prove to be inappropriate. DLSw provides the additional functionality to support these types of configurations.

A DLSw-capable device uses an enhanced bridging protocol to allow for disjointed configurations, such as an intermediate WAN. It provides this by using the following techniques:

- *Local acknowledgment.* DLSw provides a local acknowledgment to each end of the LLC2 session. This eliminates the timing problems associated with the relatively long WAN delays.
- *Local retransmission.* Retransmission of frames is provided by the DLSw device. Frames are transmitted across the WAN only once.[3] If a local acknowledgment is not received, the local DLSw device provides the retransmission; data is not retransmitted across the WAN.
- *Reduction in broadcasts.* Token ring and, especially, Netbios relies upon broadcast frames to determine the location of a destination. The transmission of these broadcasts can quickly clog the

---

[2]IBM itself recognized this as a result of its implementation. The company was a major contributor to this effort.

[3]Normal DLC retransmits can occur on the WAN, but this is only applicable to getting a frame successfully across the WAN, not to the destination station.

throughput of the WAN interface. DLSw defines several methods that reduce the transmission of broadcast frames across the WAN.

- *Multiplexed LLC sessions on to a TCP session.* A pair of DLSw devices use a pair of TCP sessions[4] for communication. Each of the TCP sessions is unidirectional. The DLSw devices multiplex link level control (LLC) sessions on to the sending TCP session and demultiplexes LLC sessions from the receiving TCP session. This reduces the number of TCP sessions between a pair of DLSw nodes that must be maintained.

In DLSw, because it uses a bridging paradigm, SDLC devices must be defined with a MAC and SAP to enable communication. This requires that values be defined for any resource that wants to utilize a DLSw-capable device. This can have implications to be discussed later in this chapter.

11.3.1 Virtual bridge

The connection formed between two DLSw devices is that of a *virtual bridge*. This "bridge" appearance encompasses the entire path between DLSw devices. If token ring devices are connecting through DLSw, this bridge appears as a virtual ring. For SDLC devices, the bridge disappears as an extension of the existing SDLC connection. Thus, if there are 10 hops between the DLSw devices, the bridge appears as a single virtual ring or as an extension of the SDLC connection.

11.3.2 Transport connection

The first event that must occur is that the DLSw devices must establish a pair of TCP sessions. The default parameters for the TCP sessions is shown in Table 11.1. In this configuration, both of the TCP session must be active for the transport connection to be usable. The DLSw node must activate the partner session if one of them fails.

[4]There are options to reduce the TCP session count to one, but this topic will not be discussed here.

TABLE 11.1 Default Parameters for DLSw TCP Sessions

Socket family	AF_INET	Internet Protocols
Socket type	SOCK_STREAM	Stream socket
Read port number	2065	
Write Port Number	2067	

11.3.3 DLSw addresses

In order to support the multiplexing feature of DLSw, it is necessary to address the transport connection between DLSw nodes, known as the *data link ID*. Table 11.2 shows the format of the data link ID. This identification is used for all logical connections between the origin and destination.

Whereas the data link ID defines a transport between two nodes in the network, these nodes can support multiple logical connections. A *circuit ID* is a 64-bit number that identifies each of the logical connections between end nodes. This identification has local significance and must be unique within a DLSw node.

An end-to-end logical connection is identified by a pair of circuit IDs. Table 11.3 shows the format of the circuit ID. Each DLSw node must keep a table of circuit and their associated data link IDs. Through the use of this table, a DLSw node can identify the logical and "physical" connections between end nodes.

11.3.4 DLSw header formats

The communication between DLSw nodes is prefixed with one of two data headers. These headers are an *information header* and a *control header*. The control header is used for all communication except for information and independent flow control messages.

Table 11.4 shows the format of the control header. This header is 72 bytes long and provides addressing for the data link and circuit IDs. Table 11.5 shows the format of the information header. This header, which is 16 bytes long, is used to identify the logical connection that is being used for a particular communication. Because the first 16 bytes of these headers are the same, the parsing of the headers is made easier.

TABLE 11.2 Format of Transport Connection Address (14 Bytes)

00	Target MAC address	
06	Origin MAC address	
12	Origin SAP	Target SAP

TABLE 11.3 Format of Circuit ID (8 Bytes)

Offset	Field
00	DLC port ID
04	Data link correlator

TABLE 11.4 Format of SSP Control Header (72 Bytes)

Offset	Field	Field
00	Version number	Header length (72)
02	Message length	
04	Remote data link correlator	
08	Remote DLC port ID	
12	Reserved field	
14	Message type	Flow control byte
16	Protocol ID	Header number
18	Reserved	
20	Largest frame size	SSP flags
22	Circuit priority	Message type (Note 1)
24	Target MAC address	
30	Origin MAC address	
36	Origin SAP	Target SAP
38	Frame direction	Reserved
40	Reserved	
42	DLC header length	
44	Origin DLC port ID	
48	Origin data link correlator	
52	Origin transport ID	
56	Target DLC port ID	
60	Target data link correlator	

TABLE 11.4 Format of SSP Control Header (72 Bytes) (*Continued*)

64	Target transport ID
68	Reserved field
70	Reserved field

TABLE 11.5 Format of SSP Information Header (16 Bytes)

00	Version number	Header length (16 bytes)
02	Message length	
04	Remote data link correlator	
08	Remote DLC port ID	
12	Reserved field	
14	Message type	Flow control byte

11.3.5 Mapping of data sessions across TCP sessions

Mapping of the data sessions to the TCP sessions between the DLSw nodes is done through the use of the circuit ID.[5] A unique circuit ID is assigned to each data session flowing between a pair of DLSw nodes that originate and terminate at the same addresses (possessing the same data link ID).

This one-for-one mapping allows for easy session outage notification and determination of the extent of an outage. It also allows for management of these multiplexed sessions by being able to determine the number of logical sessions flowing between a pair of devices in the network.

Unfortunately, this solution does not allow for easy mapping between the DLSw view of logical connections and, for example, a TP monitor view, such as CICS. The reason is that the LU names will not be readily available to the DLSw node. Because DLSw operates at the

[5]The circuit ID is cross referenced to the data link ID.

bridge, it does not have knowledge of the LUs that are associated with a given media access control/service access point (MAC/SAP) pair. The LU knowledge is above this level and few if any DLSw nodes will attempt to provide this information.

Information about NetBIOS, on the other hand, will be available. This is because a NetBIOS name is cached along with the MAC/SAP by the DLSw node. The node does this because it must forward all NetBIOS frames without regard to whether the name has been cached.[6]

11.3.6 Local spoofing

Because the SNA protocols that are being transport are optimized for a LAN environment, when they are passed across a WAN, they have a tendency to lose connectivity as the time for acknowledgments exceeds the allowable limit.

DLSw addresses this issue by providing local acknowledgment at each DLSw node. Thus, each node is responsible to provide local acknowledgment of message reception. Because the number of DLCs that are supported is limited and because each DLSw node is only responsible for its own local DLC, it is possible to provide specialized DLSw nodes that are keyed to a specific protocol. Though this technique is not embraced by all vendors, some have found this to be an economical methodology.

Hypercom, for example, provides exactly this type of specialized node. This vendor is especially well suited to this technique as a result of its hardware architecture. A Hypercom chassis encompasses a series of totally independent cards. Each has its own central processing unit (CPU), physical interface, and software. As a result, each card is responsible for a physical interface and the bus between the cards. This architecture is well suited to providing this type of optimized DLC termination.

11.3.7 DLSw node responsible for message delivery

Each DLSw node, because of local DLC acknowledgment, is responsible for the transportation of messages to the destination DLSw node. As the frame has already been acknowledged, the local DLC has no awareness that the message has not reached the actual destination.

---

[6]This is done so that the partner can become aware of the session number associated with the interface.

As a result, DLSw requires that each node become fully responsible for the delivery of the acknowledged message to the partner DLSw node and the remote node is responsible for delivery to the actual destination.

Message delivery, when using DLSw, can be viewed as consisting of three separate and distinct sections. These are:

1. *Local segment.* This is the local segment from the local end device to the DLSw node. The local DLSw node provides the DLC termination along this segment.
2. *DLSw-DLSw segment.* This is the segment across the WAN connection between the DLSw nodes. The TCP session between the nodes is the DLC for this connection. This session provides for the error recovery and message retransmission between the nodes.
3. *Remote segment.* This is the segment between the remote DLSw node and the destination device. The remote DLSw node must guarantee that the message is delivered to the actual destination device.

With the disjointed nature of the connection, problems associated with being unable to complete message delivery are inevitable. What is most difficult is how to handle the event when message nondelivery occurs. As with several portions of this architecture, the method for resolving this event is up to the particular implementation. Since this is a key portion of the design, there will be a large variation in how effectively DLSw nodes operate.

The remote segment is the most problematic element of the communication path for the DLSw architecture. A frame is transmitted across the WAN only once. As a result, once a frame has successfully traversed the WAN, it is the responsibility of the remote DLSw node to successfully pass the message to the destination device. Since this is completely isolated from the original transmission, recovery if the message cannot be delivered is a challenge because by this point both the local DLSw node and the remote DLSw node have acknowledged the frame. In order to enable any type of recovery, a private protocol must exist between the pair of DLSw nodes.[7] Even if it does exist, because the frame has already been acknowledged, there is no way to inform the originating station of the delivery failure. The only way[8] to

---

[7]Even if such a private protocol exists, recovery will be difficult because of the message acknowledgment.

[8]This is the method proscribed by both DLSw RFCs.

resynchronize the logical connection is to signal termination of the logical circuit and allow it to be reestablished or to withhold the local acknowledgment until the message has been *remotely* acknowledged.

11.3.8 Explorer and circuit start frames

RFC 1795 expanded on the circuit establishment technique developed in RFC 1434. In RFC 1795, a differentiation was made between the need to locate a resource and actually establishing the connection to that resource.[9] Thus, two types of locator frames are supported for LLC1 and LLC2 connections. These are *explorer* and *circuit setup* commands.

The explorer frame is sent to explore the topology of the DLSw interconnections. These frames do not result in an actual logical circuit to be established. Instead, they provide a DLSw node with knowledge of the network topology; i.e., where resources exist and the route to that resource.

Transmission of an explorer (_ex) frame is triggered by reception of a TEST frame, a broadcast XID, or a NetBIOS NAME_QUERY. The DLSw node either scans its cache[10] or transmits an _ex frame to look for the specified destination.

The explorer response carries the response to the originating station. This station makes a determination of which route will be utilized and a nonbroadcast XID or a SABME is directed at the destination node. The DLSw node can optionally cache the route chosen for the connection so that a broadcast search is not necessary for the next request to the same destination.

At this point, a directed circuit start (_cs) frame is transmitted. This frame results in the DLSw node actually reserving and building the control blocks that will be used while in the connected state. Figure 11.1 shows the use of the explorer and circuit start frames.

On sending the _cs frame, the DLSw enters a busy state to the originating station. This stops this station from sending any frame that cannot be completed until the circuit establishment is complete.[11]

If the location chosen by the DLSw node is cached, the next request to the same destination will not be preceded by the use of an explorer

---

[9]In RFC 1434, a circuit was established by a topology request. This resulted in circuits that were not optimal and in the establishment of circuits when a connection was not desired.

[10]A DLSw node can contain no cache; it may also not contain a MAC and/or NetBIOS name cache.

[11]In this case, this occurs when a UA frame is received from the partner DLSw node originating from the destination station.

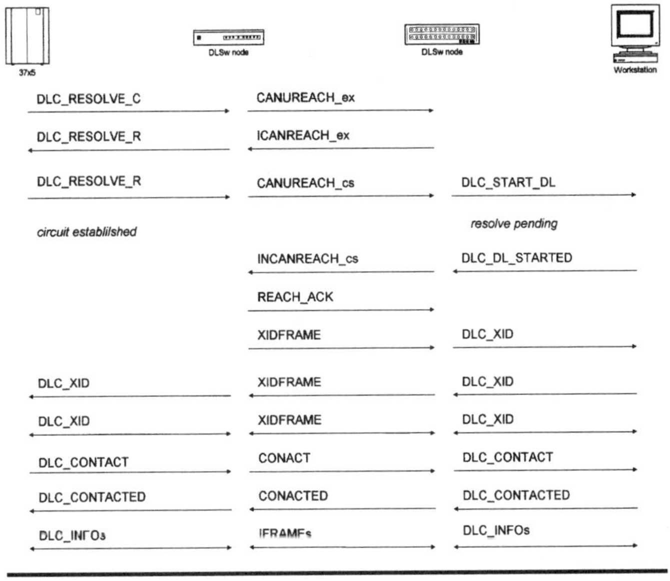

37x5	DLSw node	DLSw node	Workstation
DLC_RESOLVE_C	CANUREACH_ex		
DLC_RESOLVE_R	ICANREACH_ex		
DLC_RESOLVE_R	CANUREACH_cs	DLC_START_DL	
circuit established		*resolve pending*	
	INCANREACH_cs	DLC_DL_STARTED	
	REACH_ACK		
	XIDFRAME	DLC_XID	
DLC_XID	XIDFRAME	DLC_XID	
DLC_XID	XIDFRAME	DLC_XID	
DLC_CONTACT	CONACT	DLC_CONTACT	
DLC_CONTACTED	CONACTED	DLC_CONTACTED	
DLC_INFOs	IFRAMEs	DLC_INFOs	

Figure 11.1 DLSw startup with XID.

frame. Instead, the DLSw node will use a _cs frame to establish a circuit to the destination, without requiring the use of an _ex frame. This can greatly speed up the process of initiating a circuit to the destination. This process is abandoned only if the circuit setup fails, in which case the base process of using explorer frames is utilized.

11.3.9 Capabilities exchange

When DLSw nodes establish their TCP sessions, and optionally during normal data flow if there has been a change in an operational parameters have been modified, they exchange a set of messages that define their own capabilities. This exchange allows the session partner to gain an understanding of how to interact with its partner node.

This exchange flows within a single GDS variable as a set of LT-type[12] subfields. Figure 11.1 shows the format of the GDS datastream used for a capabilities exchange. Table 11.6 shows the set of subfields defined for use in DLSw. Because some of the subfields build upon knowledge gained from previous subfields, the following required subfields must appear in the following order:

---

[12]An LT-type field is formatted as LLTT..data..; where LL is a 2-byte length field inclusive of the length field itself, and a 2-byte type specifier that defines the data that follows.

TABLE 11.6 DLSw Capabilities Exchange Subfields

Subfield description	Hex value
Vendor ID	81
DLSw version	82
Initial pacing window	83
Version string	84
MAC address exclusivity list	85
Supported SAP list	86
TCP connections	87
NetBIOS name exclusivity list	88
MAC address list	89
NetBIOS name list	8A
Vendor context	8B
Reserved for future use	8C–CF
Vendor specific	D0–FD

1. Vendor ID
2. DLSw Version
3. Initial Pacing Window
4. Supported SAP List

The rest of the subfields can appear in any order. In addition, some of the subfields can be repeated. These include the MAC Address List and NetBIOS Name List subfields. There is no mandatory order for the repeated subfields. Thus, they may appear anywhere in the set of subfields and repeated subfields do not have to be adjacent.

11.3.10 Switch to switch protocol

Communication between DLSw nodes is provided through a switch to switch protocol (SSP). This protocol defines the requests and responses that allow for network searches for resources and for data transmission. It also allows the DLSw nodes to signal the initiation and termination of DLCs on their respective sides.

RFC 1795 includes the finite state machines (FSMs) for the initiation and termination of logical circuits, the passing of requests and responses for resource searches, and other types of control flows.

Table 11.7 shows the defined messages that make up the SSP. All of the messages use the 72-byte control header except for INFOFRAME, KEEPALIVE, and IFCM. As the rest of the frames do not normally carry any user data,[13] the length of the control header does not have a large effect on data transmission or throughput.

[13]The CAP_EXCHANGE header is one of the exceptions.

TABLE 11.7 SSP Commands

Commands	Description	Value (hex)	Flag notes
CANUREACH_ex	Can U Reach—explorer	03	SSPex
CANIREACH_cs	Can U Reach—circuit start	03	
ICANREACH_ex	I Can Reach—explorer	04	SSPex
ICANREACH_cs	I Can Reach—circuit start	04	
REACH_ACK	Reach Acknowledgment	05	
DGRMFRAME	Datagram frame	06	Note 1
XIDFRAME	XID frame	07	
CONTACT	Contact remote station	08	
CONTACTED	Contacted remote station	09	
RESTART_DL	Restart data link	10	
DL_RESTARTED	Data link restarted	11	
ENTER_BUSY	Enter busy	0C	Note 2
EXIT_BUSY	Exit busy	0D	Note 2
INFOFRAME	Information frame	0A	
HALT_DL	Halt data link	0E	
DL_HALTED	Data link halted	0F	
NETBIOS_NQ_ex	NetBIOS name query—explorer	12	SSPex
NETBIOS_NQ_cs	NetBIOS name query—circuit start	12	Note 3
NETBIOS_NR_EX	NetBIOS name recognized—explorer	13	SSPex
NETBIOS_NR_cs	NetBIOS name recognized—circuit start	13	Note 3
DATAFRAME	Data frame	14	Note 1
HALT_DL_NOACK	Halt data link with no acknowledgment	19	
NETBIOS_ANQ	NetBios add name query	1A	
NETBIOS_ANR	NetBios add name response	1B	
KEEPALIVE	Transport keep alive message	1D	Note 4
CAP_EXCHANGE	Capability exchange	20	
IFCM	Independent flow control	21	
TEST_CIRCUIT_REQ	Test circuit request	7A	
TEST_CIRCUIT_RSP	Test circuit response	7B	

Note 1: DGRMFRAME and DATAFRAME carry data as a UI frame. DGRMFRAME uses the circuit ID pair for addressing, while DATAFRAME uses the Data Link ID.

Note 2: These messages are used by older nodes that implemented RFC 1434. They are not part of the current DLSw standard.

Note 3: These messages are used by older nodes that implemented RFC 1434. They are not part of the current DLSw standard. The replacement for this message is NB_*_ex.

Note 4: A KEEPALIVE message is sent to verify the TCP session. If one is received, it can be discarded.

11.4 DLSw Design Analysis

The DLSw design provides a good foundation for the transportation of SNA data across an IP backbone. It provides for method of overcoming some of the most troubling areas when supporting this type of topology. Among the problems addressed by DLSw, according to tables within the DLSw specification, include:

- DLC timeout
- DLC acknowledgments over a WAN
- Flow and congestion control
- Broadcast control of search frames
- Source route bridging hop count

To this end, DLSw is successful in providing functionality. It has developed methods of solving these problems, while not requiring the implementation of any proprietary solutions. This is not a simple task to perform, but DLSw is able to provide it today.

11.4.1 Session timeout problem resolved

Among the problems addressed by DLSw is the timeout of sessions caused by the long latency time caused by LAN protocols crossing a WAN. The design requirement for local acknowledgment overcomes this problem by providing acknowledgment of a frame at the local interface rather than having to wait for this acknowledgment from a remote node.[14]

11.4.1.1 Analysis of session timeout. DLSw provides for the termination of the local DLC. This eliminates the timeout of the session that results from a delay in session response time. This development is a positive advance for the cause of SNA connectivity.

On the other hand, by providing a local acknowledgment before the frame has actually reached its destination, a problem arises if the frame cannot actually be delivered. In this case, there is no way of informing the originator of the message that a frame, which has already been acknowledged, has not actually reached its destination.

The mode that DLSw has taken[15] is to sever the logical connection. This breaks all of the acknowledgment chain, allows for the connection to be reestablished, and the acknowledgment chain to be restarted. Unfortunately, this is completely disruptive to the existing connection.

There are alternatives to this catastrophic loss of connectivity. Several vendors, e.g., Hypercom, allow for a delayed acknowledgment. This allows the acknowledgment to be received transparently that data was delivered to the remote destination. As has been mentioned, receipt of the above acknowledgment runs the risk of taking

---

[14]Of course, the local acknowledgment results in new problems caused by acknowledging frames that have not actually been delivered to the destination.

[15]This is the only viable method given DLSw's mode of operation with acknowledgments.

longer than the acknowledgment timer of the DLC used by the local device, which results in session outage.

The problem is one that needs an alternative methodology. Neither answer is sufficiently well behaved not to lead to some type of problem. A method of acknowledging a frame at the link level is needed that does not result in one of the layers perceiving this as acknowledging that a frame has been successfully delivered. Though proprietary methods exist for implementing this solution, no standard DLCs allow for this type of operation.

11.4.2 Flow and congestion control

Subarea SNA, APPN, and TCP have flow control on a session basis. This flow control method is based on the idea of a *send window,* which specifies the number of transmits that can be issued without requiring an acknowledgment of the transmission.

HPR supports APPN flow control, but also can utilize ARB flow control. This flow control method uses a window that is adaptive to the throughput that is being realized on both sides of the connection. Although this is an advanced flow control algorithm, the impact on throughput across a DLSw connection is minimal.

The SNA architectures also provide flow control at the link layer. Because the SNA protocols are all based on a session at a higher level, the flow control mechanism is based on a flow control at a higher level. IP, on the other hand, does not recognize the same type of flow control. This is because frames at this level are all datagrams. As such, there is no basis for flow control or recovery.

DLSw supports a flow control procedure between nodes. Though the method is not fully developed, it is a step in the direction of supporting forward and backward flow control procedures. This allows for backward flow control to be applied so that the transport between DLSw nodes is not overrun with data.

11.4.2.1 Analysis of flow control. The flow and congestion control method described in RFC 1795 is not complete. It makes assumptions about the network topology and provides limited flow control. Since the method has not been implemented nor has it been modeled for operational characteristics, the effects of the described method are unknown. It is also unknown if a congested state can actually be exited or if the flow control mechanism is sufficiently effective.

Because the receiver grants the sender approval for a specific window, it is felt that the tools are sufficiently robust for use. Since the receiver is the station with a buffer problem, allowing it to determine when flow control is required is a reasonable choice. In addition, because flow control indicators can flow on information frames, it is reasonable to expect that flow will be controlled and that deadlocks will not be a serious problem.

At the same time, analysis of this type of process is extremely complex. It is not known if a method that works under normal conditions can assume quite different characteristics when congestion actually occurs. This is why field testing and advanced modeling techniques should be utilized.

11.4.3 Broadcast reduction

DLSw specifies a method of reducing the required broadcast frames and searches throughout the network. By optionally caching information concerning destination addresses, a DLSw node can respond to a broadcast search without requiring the broadcast to be passed to every segment of the network.[16] This can greatly reduce the broadcast searches that are indicative of LAN devices using source route access.

DLSw also allows node-by-node definition by node of what SAP or NetBIOS names are (or which are not) serviced by a node. This allows the originating node to determine which DLSw nodes to interrogate if a request comes in for a specific destination.

11.4.3.1 Analysis of broadcast reduction. The design of DLSw provides several excellent steps toward the reduction of broadcasts across the connection between DLSw nodes. These steps reduce the searches of the network for a destination, so that only by specifying definition could flows be further reduced.

Although it is easy to build functionality in this area, DLSw design provides about as much reduction as possible. When caching is used,[17] along with SAP and NetBIOS name lists, broadcast searches are greatly reduced. It would appear that only predefinition would provide a greater reduction in broadcast searches, a step that no one would favor!

11.4.4 Source route hop count

DLSw creates a virtual ring between DLSw nodes. This virtual ring is sensitive to the actual hops that must be traversed when crossing the WAN between nodes.

---

[16]This assumes that the destination address has already been cached by the DLSw node. If this is not the case, the frame is broadcast to every node that has advertised that it supports the SAP and/or NetBIOS name.

[17]Although caching is an option, if would be surprising to find any implementation that does not cache IP addresses and NetBIOS names.

Depending on how your network is built, the virtual ring between DLSw nodes can provide isolation from exhausting the maximum hop within your network. This solution is implemented by using only one hop for the hop count between DLSw nodes, no matter how many hops are actually crossed along a particular path.

11.4.4.1 Analysis of hop count reduction. DLSw provides an aid to the hop count limit imposed by token ring networking. Although this solution, virtual ring, is not unique to DLSw, the designers of DLSw did recognize the problem and included this into the ultimate design. As such, this solution provides the ability to potentially extend the span of a token ring network by providing a bridge between LAN segments without incurring the hop count cost.

11.4.5 Other topics

The following topics are other areas of analysis that an user of DLSw should know about. These topics illustrate some of the design implications of DLSw.

11.4.5.1 TCP sessions. DLSw nodes open two TCP sessions to every other DLSw node within the network. When there are only a few nodes, the overhead associated with this number of TCP sessions is not of interest. But when there are hundreds or thousands of DLSw nodes, the TCP session count can exceed the capacity of even the largest node. This results in a restriction in either the size of DLSw networks or the topology of those networks.

In order to resolve this design problem, two answers have been suggested. The first is to reduce the TCP session count. If only one session is established, this potentially doubles the number of DLSw nodes that can be supported. Although this solution reduces the problem, it does not eliminate it.

The second solution is to partition the network into smaller pieces.[18] This reduces the size of the network with which most DLSw nodes have contact. By reducing the number of partner nodes, this partitioning reduces the memory usage within the node. To support communication between these smaller subnetworks, gateway nodes are used. These gateway nodes must have awareness of each of the DLSw subnetworks and establish connections between the subnetworks. It is now demanded that these gateway nodes support cross-

[18]cicso Systems has made a proposal to the AIW regarding this design point.

network communication and that they control the logical circuits used between the subnetworks.

If your network requires only communication within a subnetwork, this solution will fulfill your requirements. If, on the other hand, communication is spread among a larger set of subnetworks, this solution has not resulted in much gain. Although the overhead on a base mode within the network has been reduced, the gateway nodes must still support heavy memory usage to support cross-network communication. It is still potentially difficult for the gateway nodes to support the level of memory required by this design.

These modifications do reduce the memory utilization of DLSw, but the design still uses a large amount of memory. This memory is taken up with:

- TCP control blocks
- Data link ID control blocks
- Circuit ID control blocks
- Origin and target transport ID control blocks
- IP address cache
- NetBIOS name cache

Taken together, this adds up to a very sizable level of memory utilization. What is required is not a cosmetic correction, such as the use of one TCP session to each DLSw node, but rather a more far-reaching solution that reduces the number and size of control block usage.

11.5 Management of DLSw Network

Management of the DLSw nodes, and the logical network that is created by the architecture, has been architected to be managed by SNMP. A special MIB has been created that defines the information that can be obtained by a DLSw node.

It's interesting that this direction was taken with DLSw, because this is not the normal SNA management interface. One could easily ask the question, Why was this done for SNA data?

The reason becomes evident when you look at the nodes that will be providing the SNA transport, remembering that TCP/IP is the preferred method[19] of SNA transport between DLSw nodes. These

[19]In fact, TCP/IP is the only architected method. There has been a great deal of discussion about alternative transports, but none has been agreed upon.

TCP/IP nodes are almost always routers. SNMP is a very common management interface for IP routers. Thus, we see the reason for the choice of SNMP as a management interface.

Use of SNMP results in an open management platform, but reduces the visibility provided by SNA management tools, such as NetView. This is because NetView normally utilizes SNA management services. This architecture uses architected flows from a PU, in subarea SNA, or an architected TP, in APPN and HPR. Although some subarea SNA, APPN, and HPR nodes can respond to SNMP requests, this is not their native management interface. Likewise, although NetView has tools that enable SNMP requests and responses to be understood, this is an alternative interface that is *not* the norm.

11.6 Summary

DLSw was created as a standard method of allowing SNA and NetBIOS data to be transported across a nonnative network. An IP transport network is the foundation of the backbone between DLSw nodes.

This type of network architecture is created for several reasons. Among these are the inexpensive cost of TCP/IP components. There has been such a growth in this segment of the networking arena that these components have become commodities.

The development of client/serve products that have been built on inexpensive platforms that often use TCP/IP as their native transport has increased the development of IP-based networks.

The objective of DLSw was to develop a *standard* method of allowing SNA and NetBIOS to be transported across IP-based networks. Although several vendors had developed proprietary methods, interoperability was all but impossible.

For this reason, IBM developed the first specification for DLSw. This was the informational RFC, RFC 1434. This specified a standard method of transporting SNA data across an IP network.

As the demands for this type of configuration are large, the design had to overcome several potential problems. Among these were:

- Session timeouts caused by the long delay in acknowledgments arriving at an origin.
- The retransmission to a destination was done locally, rather than requiring retransmission across the WAN interface.
- Reducing the broadcasts that are carried across relatively slow WAN interfaces.

- The multiplexing of multiple LLC sessions across a single TCP/IP session.

The interconnection between DLSw nodes is that of a *virtual bridge*. This bridged configuration allows the interface between a pair of DLSw nodes to be viewed as a single hop. The elimination of hops is especially important for token ring networks.

Addressing between DLSw nodes is complex. The first level of addressing is of the logical connection between a pair of DLSw nodes. This address is called a data link ID. This allows the logical connection to be identified.

The sessions flowing across the data link ID also must be individually addressed so that they can be properly demultiplexed and routed at the remote end. Each of these sessions is identified by a circuit ID. This 64-bit identifier is paired with the circuit ID in the remote DLSw node to define an end-to-end session.

DLSw has defined two types of headers. The first is the control header. This 72-byte long header allows for addressing the data link and circuit identifiers. This header is used for the switch–switch protocol (SSP) between DLSw nodes.

The second header is the informational header. This 16-byte header allows session data to be multiplexed. As these 16 bytes directly match to the first 16 bytes of the control header, parsing of these messages is made easier.

One of the most difficult tasks for a DLSw node is caused by the fact that all sessions between nodes outside of the IP backbone network are locally terminated. In addition, these interfaces have local acknowledgment. As a result, it is possible that a local resource obtained a local acknowledgment for a data transmission, but that the message cannot actually be delivered. This can cause a great deal of difficulty based on the fact that a link acknowledgment has already been received. Several vendors have looked at this potential problem and have developed a proprietary method of overcoming this problem.

IBM and several vendors developed a venue for discussion of DLSw as a part of the APPN Implementors Workshop (AIW). The related interest group (RIG) that has been an outgrowth has created a newer RFC. RFC 1795 extends the capability of DLSw.

The largest two additions to the architecture are the split between explorer and session establishment tasks and the development of a flow control algorithm.

The explorer frames are used to extend the topology knowledge of a DLSw node. Only after this topology has been obtained is a circuit start frame sent directly to the destination. By splitting these features up, no inappropriate circuits are created.

RFC 1795 also added a capability exchange between DLSw nodes upon attempting to establish TCP sessions. This capability exchange is similar to the same logic in APPN. It allows DLSw nodes to adjust dynamically to changes in the network and the nodes within that network.

When looked at in total, DLSw does an effective job of creating a standard method of interconnecting SNA and NetBIOS resources across IP networks. Although there are still many areas that need to be developed further, DLSw has made a mark in the industry and has pointed the direction for other methods of providing interconnection across nonnative networks.

Chapter

12

Software Products for SNA and IP Environments

I have discussed several methods of integrating subarea SNA, APPN, HPR, and TCP/IP on a large scale. The methods analyzed are meant for the integration of these network architectures so that information can be freely intermixed and, optimally, a single physical interface can carry all of the data traffic.

This chapter will review methods that are largely used for a smaller-scale integration task. The result of implementing the methods described in this chapter can provide a single physical interface, but they all exhibit some type of limitation.

Often these interfaces are used either to communicate with a single destination or for a particular, short-term purpose. These interfaces often require more definition requirements than the more generalized interfaces[1] described in other chapters.

12.1 AnyNet

AnyNet is a product developed by IBM to allow API portability. The aim of this product is to separate the application layer from the network architecture that is used.

As I stated earlier in this book, the application layer (API) is largely dictated by the network architecture that is used. For example, if an SNA network is being used, it is common for the applications to be based on IBM technology, e.g., a CICS transaction system. Likewise,

[1]Since some of the general interfaces have a high definition requirement, this statement may be hard to believe. But these interfaces are meant for a solution to a particular requirement, rather than a large-scale integration task.

if a TCP/IP network is being used, it is quite common for an interface such as *sockets* to be utilized.

AnyNet is part of the IBM Networking Blueprint. This plan defines a method of separating the application from the network architecture portion of the network design. This design, known as Multiple Protocol Transport Networking (MPTN), defines how applications can operate on a network architecture that is not native.

This design allows, for example, CICS applications to communicate with IP workstations and a socket application to communicate with an SNA 3270 device. This ability to intermix applications and end devices eliminates the requirement to match these endpoints of the network.

AnyNet provides a translation interface between these interfaces that allows this type of interaction to be provided. Figure 12.1 depicts where this layer resides and gives some idea of how it provides its

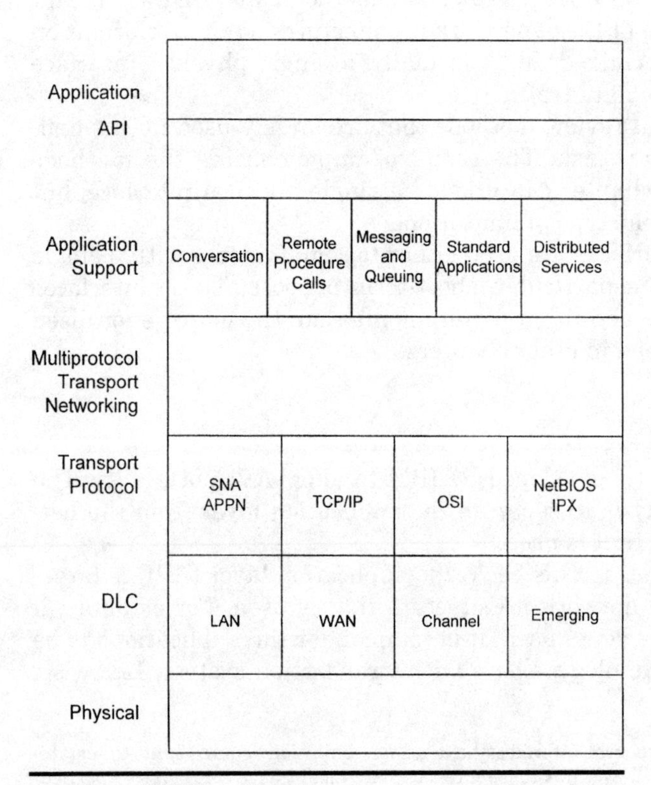

Figure 12.1 MPTN configuration.

services. AnyNet provides a translation between the application services and network protocols. In some cases, this translation layer is a straightforward mapping of services. An example of this is address mapping between subarea SNA and IP. Although the addresses are very different in size and meaning, it is possible to create a service that provides this address mapping.

In other cases, a service does not exist and must be simulated by the translation layer. An example of this is the IP datagram when moving to one of the SNA architectures. Because subarea SNA, APPN, and HPR have no equivalent feature, it must be simulated by the translation layer.

12.1.1 Interconnection types supported by AnyNet

AnyNet is actually a group of products that provide one piece of a total solution for interconnection. Each of these provides a solution to a single type of interconnection requirement. These solutions can be categorized into the type of application that is being supported and the transport that is being used.

The transports include TCP/IP and subarea SNA. The type of applications include APPC, IPX, NetBEUI, SNA, and sockets. Together, these interfaces provide a method of allowing similar application layers to communicate, even if the transports are different.

For example, an SNA application using SNA transports can communicate with an SNA application that is using a TCP/IP transport. One of the AnyNet SNA gateways can provide this communication by compensating for the incompatibilities of the two transports.

Figure 12.2 shows how this connectivity is provided. You can see that the SNA session from the application on Host A passes through the AnyNet gateway and continues across the IP network to the SNA application operating on Host B.

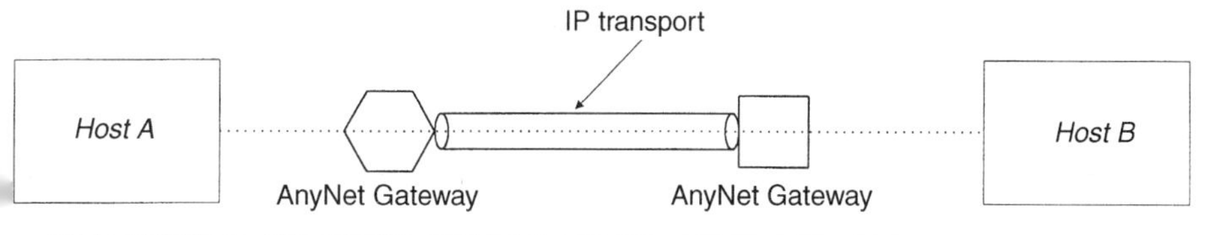

Figure 12.2 AnyNet gateway.

12.1.2 AnyNet limitations

Although these applications may be accessed across nonnative protocols, AnyNet provides only limited mapping for communication between, for example, a socket application and an SNA application.

Interconnection does imply that different network protocols are being used, but it often means that different application layers must be supported. AnyNet only addresses the first of these requirements. AnyNet does allow the same type of applications to communicate across a diverse set of transports.

The mappings that AnyNet supports does not allow diverse application types to communicate. This results in a large limitation in the use of AnyNet for connectivity of diverse networks.[2]

It is also questionable if there is a large requirement to support applications operating on a nonnative environment. Although there are times when this is a real need, the frequency of this is not large. This is becoming especially true as applications are becoming available for different environments.

When this need does arise and the environment is supported, AnyNet is a possible solution. This solution is good for a temporary need or one that does not require the highest throughput. Because of the overhead involved in the gateway functions of AnyNet, the throughput is restricted.[3]

12.2 TCP/IP for SNA Environments

Although SNA is not able to freely communicate with an IP network, it is possible to use IP communications within an SNA environment. What occurs is that a specific communications path uses an IP stack instead of SNA. This is similar to when you have SNA and bisynchronous devices communicating with a host.

IP for SNA communication interfaces have two different methods of establishing communication. These are:

1. *An addendum to normal SNA communications.* When using this mode, the TCP/IP stack is placed as an adjunct to the SNA networking. In order to provide communication in this mode, the IP stack is configured as a translation from IP to SNA. Figure 12.3 shows how this operates.

---

[2]Of course, there are few products or standards that provide this gateway facility.

[3]IBM is working hard to improve this property of AnyNet. The product has improved over the last few years and development shows that more improvement will be forthcoming.

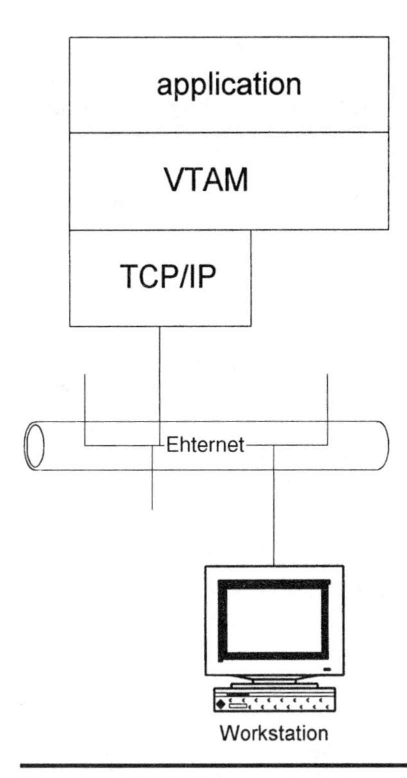

Figure 12.3 TCP/IP communication as adjunct to VTAM.

2. *A native communication interface.* In this mode, the IP stack becomes the native communication interface.[4] The communications are natively IP from the application, to the network, and to the remote destination.

Each of these interfaces provides a different level of integration and capability. The choice of which direction to take must be made because of the level of integration of TCP/IP that you need. Either method allows both TCP to be driven from the host. They also both allow access to standard TCP/IP tools, such as FTP, telnet, and tn3270. They also support network management through SNMP.

What changes is that if TCP/IP is provided by an extension to an SNA interface, the SNA stack operates as well as the TCP/IP. This

[4]Although this may seem like it should not fit within an IP on SNA interface, I describe it here because the environment is normally an SNA one. An example of this is when an IP stack is the native interface on an IBM mainframe.

results, naturally, in a higher overhead on the system. In addition, the SNA interface must have an interface to the defined TCP/IP stack. Because of the differences between SNA and IP, this configuration can be less than straightforward.[5]

For example, one of the means of providing this interface is through the IBM program product, TCP/IP for MVS.[6] When using this product, it is necessary to define the TCP/IP stack as an SNA application. It is through this indirect interface that the TCP stack is able to operate as a destination to an SNA application.

The TCP/IP stack also becomes available to native IP applications. These applications can use any of the standard programming APIs, such as socket or RPC. These native applications can communicate with both SNA and IP resources. Internet Protocol resources are accessed through either the IP stack installed on the host system, if the resource is host-based, or through the IP stack directly, if the resource is IP-based.

This type of stack supports most standard IP applications. These include ICMP applications, such as ping, UDP applications, such as trivial file transfer protocol (tftp), and TCP applications, such as ftp.

12.2.1 TCP/IP for MVS and VM

TCP/IP for MVS and VM operates as a separate address space, in MVS, or virtual machine, in VM. Communication to the TCP/IP network is provided by passing data through the TCP/IP address space. It is there that address conversion is done from the SNA address to an IP address. This conversion can be done in various ways that range from a direct, defined mapping through the use of an algorithm that allows for the transformation.

All standard IP applications are supported including ftp, telnet, and tn3270 through the TCP/IP stack. This implementation also supports an SMTP gateway between the host and the standard SMTP mail interface.[7]

These products also allow support for SNMP network management. These products provide an interface that allows a NetView operator to solicit and obtain network management information from an

---

[5]Of course, many of the definitions for SNA networks are difficult and confusing.

[6]There are also versions of TCP/IP for the VM and DOS/VSE operating systems.

[7]The TCP for VM product also supports an interface between the RSCS mail and SMTP. This allows most VM mail to be addressed to an IP network, as well as incoming mail.

SNMP MIB format. This interface is implemented through the use of a database, MIBDESC.DATA, and the SNMP Query Engine process.

SNMP requests and responses pass through this translation process. A NetView operator can obtain MIB data by entering a short form of the standard MIB variable. The MIBDESC.DATA database is utilized to translate the short name into the ASN.1 variable. This information is then extracted from the MIB and returned to the requester in the short-form status. Likewise, asynchronous alarms are converted into short-form status and passed to the NetView console as an alarm.

12.2.2 Native TCP/IP stacks on a host processor

It is also possible to utilize a native TCP/IP stack on a host processor. These implementations provide an entirely different environment than the normal operating system/VTAM pair.

An example of this type of implementation is the UTS System provided by Amdahl. This system provides a UNIX operating system on an ES/9000 processor. The standard UNIX system comes with all of the normal UNIX commands and utilities, processes, and communication interfaces, such as TCP/IP.

The UTS System is an interesting mix of UNIX processing with support for many of the normal SNA equipment, such as 3172, 4745, and the Enterprise System Connection Architecture (ESCON) channel. As result, UTS provides a mixed environment that takes the place of a VM or MVS operating system on an ES/9000 processor, while also allowing support for the UNIX commands, utilities, and communication interfaces.

Although the UTS System is not an SNA implementation, it does support IBM and IBM plug-compatible equipment. Unfortunately, this implementation does not include remnants of the key host SNA component, VTAM. This means that this environment is no more SNA-ish than any other UNIX system. Thus, it operates on equipment that is normally seen as only operating for SNA networks, but in fact the situation is far from this.

12.3 Software Products for IP Environments

In an effort to allow communication with the pervasive SNA networks, tools for communication have been developed within the IP network environments. These tools allow communication to both SNA resources from an IP network and IP resources from an SNA network. These tools are not directed at creating large integrated networks, but support the ability of a single nonnative communication interface.

Examples of these tools include tn3270 and SNA stacks on IP hosts. In the case of the former, a single non-SNA terminal is allowed access to full-screen, 3270 applications.

At the other end of the spectrum are SNA stacks for IP hosts. These stacks allow the IP host to cross-communicate with resources in an SNA network. The SNA stack is utilized to allow access for both end user devices and program-to-program communication. These stacks almost always allow downstream SNA resources to be supported by the IP host. For example, 3270 devices can often directly communicate with the IP host. In addition, these devices can pass through the IP host and access their normal session endpoints.

12.3.1 tn3270

One of the most utilized tools for this communication interface is tn3270. This tool is an extension of the normal telnet process available on IP hosts. In addition to the normal telnet interface, tn3270 provides support for access by non-3270 terminals to full-screen, 3270 applications.

This tool can be started by a user utilizing, normally, an ASCII dumb terminal on an IP host. The tn3270 tool is implemented to provide the image of a full-screen 3270 to both the connected host and to the communicating terminal.

12.3.2 SNA atacks for IP implementations

Over the years, many IP configurations have required access to SNA equipment. Often the IP equipment was installed for a specific purpose. It is quite common for a discussion to occur about how "there will be no need to connect to the glass house." Although this was an optimistic beginning, we all know how long this statement lasts—often, not even until the equipment was installed; sometimes not even until the equipment arrived.

A connection is made to allow data to be transported between the IP and centralized mainframe. This access provides bidirectional transport.

In order to support this access, vendors of IP equipment developed an SNA interface. This access initially supported an SNA interface to a subarea SNA host. Often, this was provided as an emulated 3270 cluster controller (3x74). This type of access provided a clean interface to the host, while reducing the development requirements.

Over the years, this interface has been expanded to support peer services, such as APPC and APPN. These interfaces provide better throughput and a more flexible access.

(The rest of this section will be largely directed at the SNA services available in AIX on the RS/6000. I do this only in order to keep the discussion to a manageable size, not because of the strength or weakness of this implementation.)

AIX SNA Services/6000 allows the SNA transport to be accessed by an IP resource. This product consists of:

- System resource controller
- Library subroutines
- Extensions to standard I/O subroutines
- AIX system subroutine calls

The SNA system resource manager coordinates the allocation of SNA resources to requestors on the AIX system. The resources managed by this component include:

- The starting and stopping of SNA connections
- PU and LU services
- Status reporting
- Session services
- Security

As with all SNA networks, it is necessary to define the type of resources that are utilized within the network that interfaces with the AIX equipment. This includes the PU and LUs for each resource communicating with the SNA Services/6000 component of AIX. The inclusion of these components allows the SNA Services component to understand the topology of the network.

This component supports communication with various types of serial and LAN attachments. Among these are:

- RS-232
- RS-422
- X.21 SDLC
- X.25 LAPB/QLLC
- V.35 SDLC
- Ethernet
- Token ring

The SNA Services/6000 component operates as either a LEN or APPN node to a central host or to peripherally attached devices. This type of support allows configurations that include attachment to a host, such as an ES/9000, or to other peer nodes, such as other AIX systems that are also running the SNA Services/6000 component.

Figure 12.4 shows both of these types of connections. This figure also shows how you can position one RS/6000 running SNA Services/6000 as a gateway to a network of RS/6000 machines. The gateway feature allows a designated machine to operate as a gateway to a set of distributed nodes. All communication to node downline from the gateway passes through the gateway node.

SNA Services/6000 supports LU types 0, 1, 2, 3, and 6.2. The first four LU types are normal subarea SNA node types that reflect communication to different types of resources.

Support for LU 6.2 (APPC) allows applications on the AIX machine to communicate, via an SNA connection, to another peer node.[8] Both mapped and basic APPC verbs are supported. This allows for all APPC interfaces to be supported by the SNA Services/6000 product.

12.3.3 Communication with a subarea SNA resource

Subarea SNA resources can communicate with the SNA stack on the IP implementation. Because the SNA services on the IP host utilize a standard interface, it should be expected that it can communicate successfully with other subarea SNA resources.

This communication can support communication with both a subarea SNA host, such as an ES/9000 operating with MVS and VTAM, and dependent LUs downstream of the IP/SNA host, such as 3270 terminals.

These modes of operation allow most types of SNA communication. Support for both uplink and downlink connections are supported through this interface. Integration of the subarea SNA resources is provided by this interface.

---

[8]Although an LU 6.2 session can terminate at the glass house host, APPC is a method of communication with peer nodes. The fact that the "peer" is a mainframe is not of material significance to the communication.

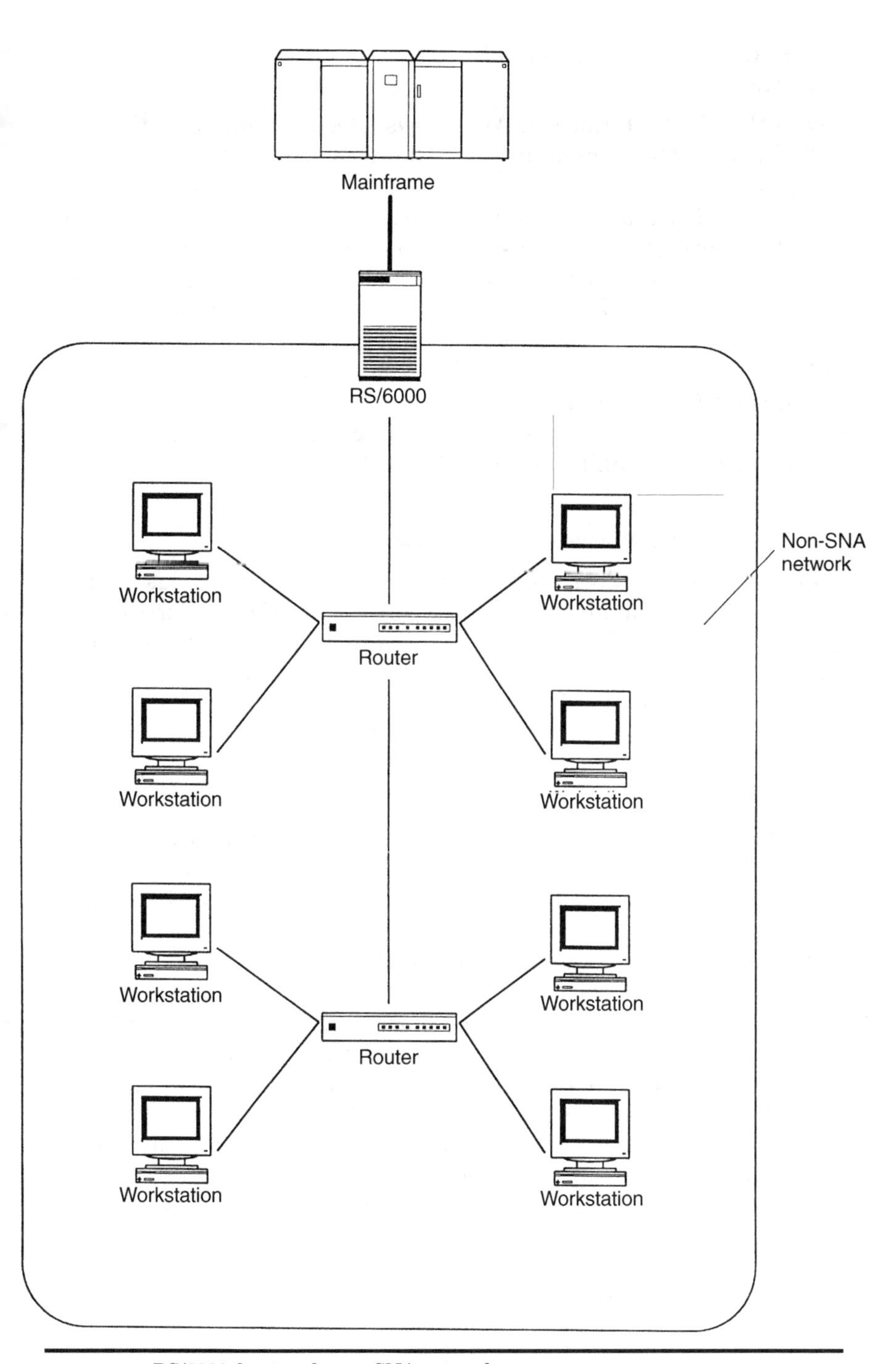

Figure 12.4 RS/6000 front-ends non-SNA network.

12.3.4 Communication with APPN resources

Use of this SNA interface for AIX allows effective communication with APPN nodes. This communication allows connection to both EN and NN resources.

The inclusion of an APPC interface expands the use of this SNA software to allow customized access to LU 6.2 resources. Since LU 6.2 is the preferred method to support SNA distributed applications, this support provides extensive use of SNA as an application access protocol.

SNA Services/6000 supports communication with an SNA host as either a LEN node or as an APPN resource. In either case, the SNA Services/6000 appears as a PU 2.1 node to the host. This type of access allows support for peer-to-peer networking that allows the IP host to provide close communication with the SNA destination host.

12.3.5 Communication with HPR resources

SNA Services/6000 does not support HPR at the time of this writing. As a result, it is unable to communicate with HPR resources as an HPR-capable node. Instead, HPR operates as a base-APPN node to those resources. In this mode, communication is fully supported, but only in the form of APPN nodes and not as HPR.

12.4 Summary

The solutions for intercommunication presented in this chapter exhibit limitations that do not make them a good fit for a large-scale implementation. Instead, these are solutions that fulfill the requirements of smaller-scale implementation.

AnyNet is a group of offerings from IBM that allow communication between application layers that do not operate on their native link layer. These offerings utilize the IBM architecture, known as MPTN.[9]

The purpose of this architecture is to isolate the application layer from the link layer. By providing this isolating interface, applications are able to be transported on nonnative link interfaces.

The isolation layer provides a translation and compensation function to both the link and application layers. These functions are necessary because of the fundamental differences between these layers.

---

[9] This architecture is laid out in the IBM Networking Blueprint.

For example, the address resolution methods for SNA and TCP are sufficiently different that direct communication is not possible. A translation facility is needed between these interfaces so that communication is possible.

AnyNet provides this translation function to compensate for differences between the layers and protocols. In some cases, the translation is done by defining a static translation. In other cases, it is possible to utilize an algorithm to provide the translation.

Although AnyNet does allow equivalent application layers to communicate on nonnative link layers, it does not provide a true integration of protocols across the nonnative link layer. Instead, the nonnative application layer is encapsulated within the link layer. For example, SNA application traffic is encapsulated within a socket interface that is specific to the AnyNet application. A well-known port is utilized by AnyNet when transporting data.

It is also possible to utilize a TCP protocol stack on an SNA platform. Examples of this includes the TCP/IP for MVS or TCP/IP for VM products. These program products support communication to IP-based applications from an MVS or VM environment.

In this case, the TCP/IP stack operates in a separate address space or virtual machine from VTAM. VTAM has definition that allow it to differentiate from the normal flow and pass the request to the TCP/IP stack. From there, the request is translated into a native IP frame and transmitted out.

The associated problem with this interface is one of the level of definition that is required, processing overhead of the additional path length, and the complexity of the interfaces utilized.

The definitions that are required to support this interface can be extensive. This includes the definition of IP address to SNA network address translation functions. As was true of AnyNet, this definition can be done either by definition or by an algorithm. The biggest differential is that in this case, the translation must support VTAM and the TCP/IP stack so that the request/response is properly routed within the network.

The additional overhead processing of the TCP/IP address space and internal routing has been greatly reduced for each release of the interface, but it is still an area that cannot be eliminated. There is also additional overhead on the host processor no matter how streamlined the code. Although this overhead should not be sufficiently high to cause concern for most installations, it is an area that must be addressed and understood.

The complexity of the interface has also been attacked over the last number of releases, but most SNA system programmers are not as knowledgeable regarding IP requirements and the definition of the interface is foreign to all. A good understanding of the interface, how it operates, how it interfaces to VTAM, and the impact of definition must be tackled before this interface can be operated efficiently. This is a large requirement that does not often fall within the immediate requirements. As a result, the interface is often operated without this knowledge and problems result.

It is also possible to operate IP as the native transport protocol on S/390 equipment. In this case, a non-SNA environment is installed on the S/390 equipment. An example of this is Amdahl's UTS system. This is a native UNIX implementation that operates on IBM mainframes and PCM hardware. To the application and communication interfaces, this is a UNIX operating system that operates an a platform that normally does not support this mode of processing.

This implementation provides for a native IP transport, but it eliminates the SNA transport, except as a nonnative interface. The solution, for obvious reasons, is not one for the questions raised by an integration effort. It merely transfers the problem of how to operate an IP network on an SNA platform to the problem of operating an SNA network on an IP platform.

Continuing our discussion of this area, there are SNA stacks that are built to provide connectivity on UNIX/IP platforms. This solution allows SNA communication to be supported by a UNIX platform, which uses IP as its native communication interface. An example of this is IBM's SNA Service/6000. This is an AIX-based product that adds an SNA stack to the native IP interface supported by the AIX platform.

This product supports PU 2.1 communication. At a minimum, this requires a LEN interface to be implemented by the host. Its most optimal configuration is as an APPN node to a host-supported APPN network.

Various LU types are supported, but the PU 2.1/APPN appearance makes an LU 6.2 session the most natural one for the software. This support allows the AIX platform to communicate with an SNA host through SNA communications.

This platform also supports communication to downstream SNA resources. Among the resources supported are 3270 terminals and LU 6.2 sessions. SNA Services/6000 allows these downstream resources to be passed to the uplink connected to the host.

Although this interface can support connectivity to subarea SNA destinations, it is architected to connect to APPN destinations. This is because it implements an APPN appearance. As a result, this interface is the best method of connecting to upstream hosts.

SNA Services/6000 can also support connection to HPR-capable hosts, but only as an APPN node. This product does not support the HPR tower to APPN. As a result, it cannot implement the additional processing required of an HPR node.

Chapter

13

Integration Analysis

We have looked at many methods of integrating subarea SNA, APPN, HPR, and TCP/IP. Each of these methods provides a different level of integration. In some cases, the integration is done at the network level. Others provide access for application-level access. All of these interfaces have in common that they attempt to connect network interfaces that are not naturally able to communicate.[1]

13.1 Shared Access Path

The first integration methods investigated were physical interfaces that allowed multiple datastreams to utilize a common communication path. These types of interfaces allow multiple datastreams to use a common path and provide support for a reduced-cost path by sharing the physical media with multiple protocols. This can reduce the number of physical connections necessary to provide connectivity and physical integration of multiple protocols. Figure 13.1 shows how a network can use less physical paths.

A cost reduction in connectivity is a positive outcome of this type of integration. The cost reduction can be seen as a tangible outcome of the integration of multiple protocols.[2] This cost reduction is a very strong benefit deriving from the creation of a shared access facility. This access facility provides a multiplexed set of isolated datastreams to a destination (or set of destinations).

[1]An exception is APPN and HPR nodes, which can natively communicate.

[2]An outcome that can be seen quickly by management as a cost savings can be a distinct advantage to your career.

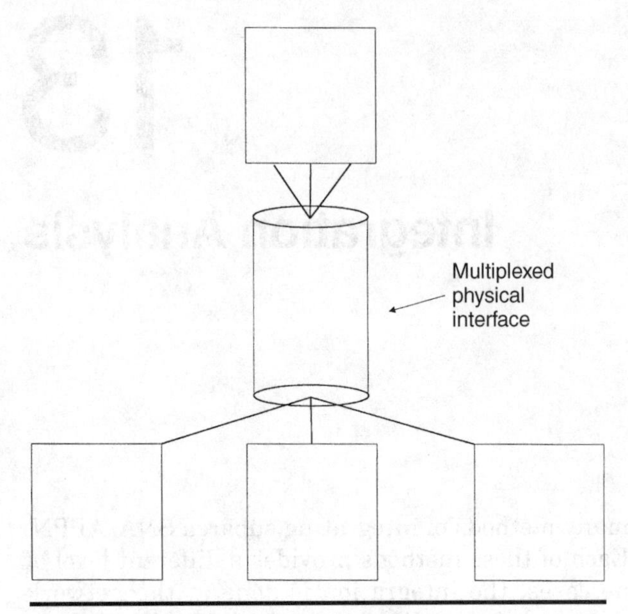

Figure 13.1 Multiplexed physical interface.

If the multiplexing is performed correctly, throughput on the shared facility can approach that of the isolated interfaces. For example, if frequency-division multiplexing is being done on the shared access link, it is feasible for each channel to operate at close to normal capacity. Figure 13.2 shows this type of multiplexing and how it can support close to normal capacity for each channel.

One of the problems with this type of multiplexing is that there are no standards that have been agreed upon so that hardware vendors can be mixed and matched within the network. Instead, each vendor has developed its own methodology that is often incompatible with another vendors. This results in a strong requirement to utilize a single vendor within the network. Although this is a boon to the single vendor, this type of solution is becoming more and more frowned upon by companies.

13.1.1 Software multiplexing techniques

Software multiplexing has developed several standardized multiplexing techniques. Because standards have been developed, it is possible to use a diverse set of vendors in the network. This ensures that a customer is not locked into a single-vendor solution. There are several standardized methods of providing software multiplexing. Among the choices are DLSw and RFC 1490.

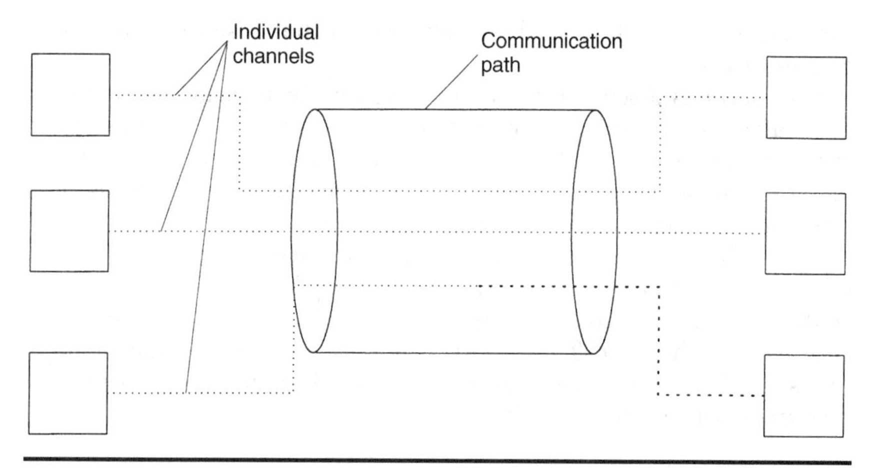

Figure 13.2 Multiplexed physical channel.

These software multiplexing techniques allow multiple protocols to traverse a reduced number of physical communication paths. The means to this end is varied between the methods, but each has its own strengths and weaknesses. As these were discussed in earlier chapters, I will not redo that work here. At the same time, a review of the type of service that they can provide is appropriate here.

Just the act of creating an integrated transport provides advantages. These advantages include the gathered information that accompanies any integration effort. This gathered information must include the consolidation of an organization's diverse pieces into a cohesive whole. This information can then be used to coordinate the diverse entities within the consistent direction and plan.

The creation of a common transport can bring an organization's diverse groups together so that they can be organized and managed. Groups not only learn what other groups have planned, but they also learn what is already available. It is possible, after this learning phase, to coordinate the plans of different groups.

I have seen many instances where different groups learn that the information that they want is already available. Often this consolidation effort allows groups that have historically worked in opposition to join forces and direct their efforts toward a consolidated goal.

It should be understood that the creation of a shared communication path does not provide true integration of resources. By this, I mean that at the application level, which is where data originates and is destined, different protocols cannot be intermixed. For example, an SNA application cannot communicate with a TCP/IP resource. Not only must the data be allowed to be transported together, but also the diverse protocols must be integrated so that resources can cross-communicate.

The methods discussed do not provide any translation of the protocol that is necessary to allow mixed resources to communicate. Each of the protocols is as distinct as before; the difference is that they can use a common transport. Without this translation aspect, cross-communication is impossible.

While the goal of creating a common transport has been met, the design point of cross-communication has not been fulfilled with this configuration. You still have distinct networks that cannot communicate with each other. These networks now have a common mode of transport, but it does not allow an integrated connection between any resources within the network.

13.2 Session Control

All of the network protocols discussed in this book are session oriented. This fact provides assurances to communication endpoints (users and applications) that the network will not only attempt to deliver their message, but will also potentially attempt to transmit multiple messages and will give feedback if the transmission was not successful. The session endpoints communicate between themselves to determine the current status, to be aware of what messages have been sent and received, and to provide methods of retrying a transmission if the session partner notes any discrepancy.

While the network protocols are all session-oriented, they do not establish or maintain these connections in the same way. As a result, direct cross-communication is not possible between these protocols. A method is needed to provide either direct cross-communication or a method of allowing the status of one connection to influence the operation of another connection.

Gateways can be created to fulfill this requirement. The gateway translates the network protocols so that resources that understand a different protocol can communicate with a resource that uses a different protocol. Figure 13.3 shows how this might be configured.

This figure shows two sessions that terminate at Gateway A. Although there is no direct session between Subarea B and TCP C, communication is possible by Gateway A because a gateway between the two sessions has been provided.

Information is taken from the session with Subarea B and transmitted to the session with TCP C. Each of the sessions is maintained by ensuring that the network protocol of each session is properly pre-

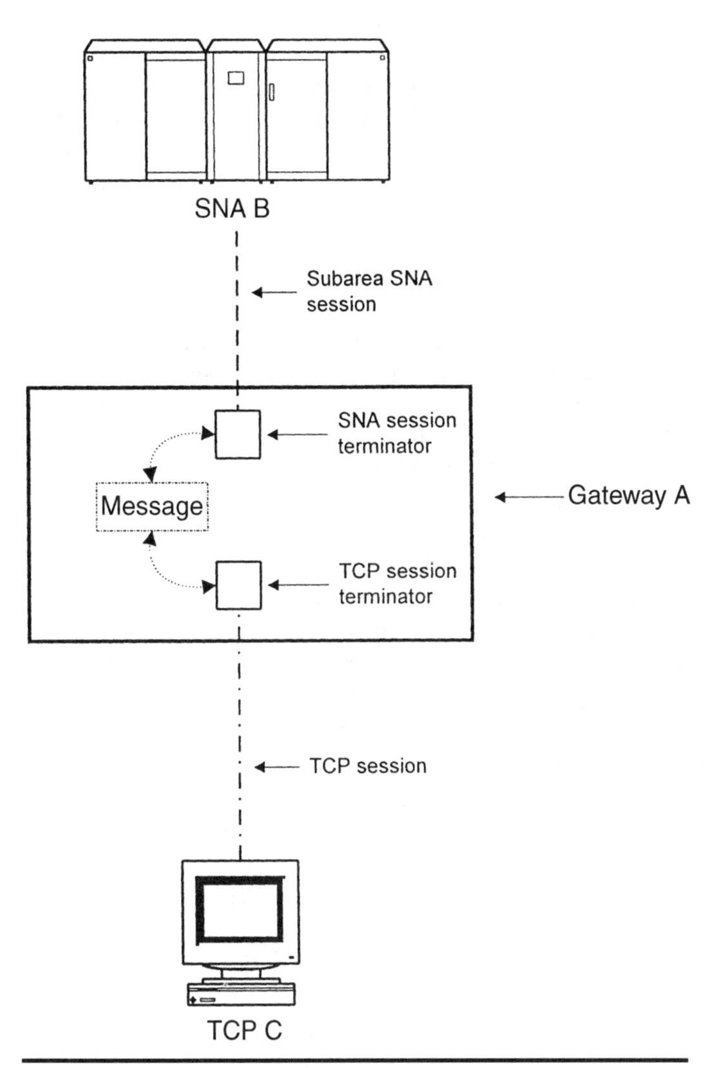

Figure 13.3 Protocol gateway.

served. This requires the gateway product to support both protocols correctly. It is not easy to find a gateway that can perform this task!

Information can now be passed from one session to the other to create the illusion of a contiguous, end-to-end session. As you can see, the TCP and subarea SNA sessions are terminated within Gateway A. The SNA and TCP sessions are maintained by managing each protocol stack to preserve the session.

13.2.1 Keeping session status accurate

One of the most troublesome problems raised by this type of attachment is keeping the status of the two sessions consistent. Because two discontinuous sessions are involved, it is common for the status of one session to not be reflected by the other session. When this situation arises, it becomes difficult to consistently determine the status of the end-to-end connection.

When one of the sessions becomes unavailable, it may not be efficient to immediately reflect this status in the other session. For example, when one session becomes unavailable, it is usually better to wait a period of time before deactivating the other session. The second session is kept active for a short period of time to determine if the first session will become active again. This procedure eliminates the overhead in tearing the second session down, only to activate it again. This deactivation/activation sequence takes a considerable amount of network bandwidth and processing overhead on each of the endpoint processors. By waiting a short period of time, this overhead is eliminated, while not negatively impacting network connectivity. Although there is a short period of time when one of the two sessions is unavailable, by properly managing network traffic and implementing a queue of requests it is possible to maintain the illusion of a contiguous, end-to-end session without any loss of data resulting.

The balancing act that must be fulfilled in performing this task is to determine the optimal time to wait until the other session is deactivated. Time is only one factor that must be taken into consideration. Others that may have an impact include:

- The size of the waiting queue
- A definition of the importance of keeping the two sessions synchronized
- The complexity of the session startup for a particular processor or special considerations

As you can see, this type or service is not a simple effort. Great care must be taken to ensure that the virtual session is maintained. It is simple to fail to maintain the illusion of a contiguous end-to-end session correctly, which may result in the loss of data.

13.2.2 Value of the gateway

This type of gateway allows applications that are built for different interfaces to communicate. This communication extends further than any other type of intercommunication because it allows different application types to communicate. All other cross-communication schemes only allow like applications to communicate.

The difficulty lies in determining a common form for data presentation. There are several directions to take, but some make more sense than others.

For example, Hypercom Network Systems has a unique set of gateway products that allow communication between a diverse set of protocols and a TCP socket resource. This group of products is called Legacy TCP Gateways. All of these products have a common presentation to the socket application. The basic configuration for this interface is shown in Fig. 13.4.

One of the keys to the socket application is to maintain an accurate view of the downstream network topology. For example, when applicable, each drop of a multidrop configuration is addressed as a separate socket. This allows the TCP application to maintain the status of the downlinks and permits direct addressing of each drop on the downstream line.

Each of these gateway interfaces terminates the protocol at the downlink and passes the user data portion of the frame to a TCP socket interface to its destination. Figure 13.5 shows the movement of the user data from the downlink frame to the uplink TCP/IP frame.

This method eliminates any encapsulation of data. The user data portion of the downlink frame is moved to the TCP/IP uplink frame. No portion of either frame is encapsulated during the transport. Instead, each side is handled in its native protocol. The user data is seen as a transparent resource that is shared between the two interfaces. It can be picked up out of the downlink frame and deposited into the TCP frame without modification.[3] The response is returned within a TCP frame, extracted, and put into an outgoing frame to the downlink connection. The only dilemma arises if the downlink sustains the same throughput rate as the TCP interface.

Although this is not an actual limiting factor, these gateways often use downlink protocols that are not designed for high throughput. Examples of the gateways that are available include:

- SNA boundary interfaces, such as LU2 and LU0
- 3270 and 2780/3780 BSC

[3]Some of the gateways do allow for a conversion of the character code that is used between ASCII and EBCDIC. This conversion is a basic tool available to the downlink frame and is not specific to the gateway.

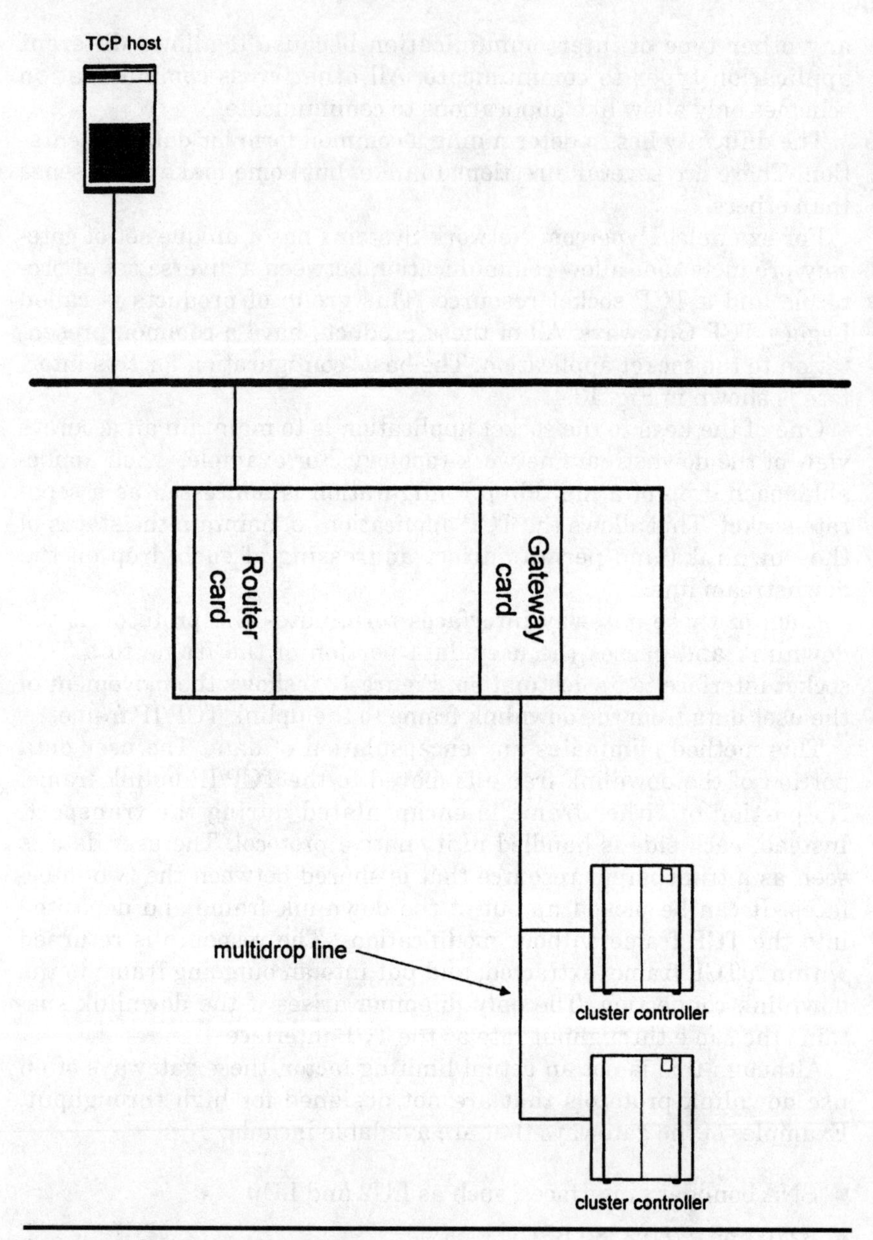

Figure 13.4 Gateway configuration.

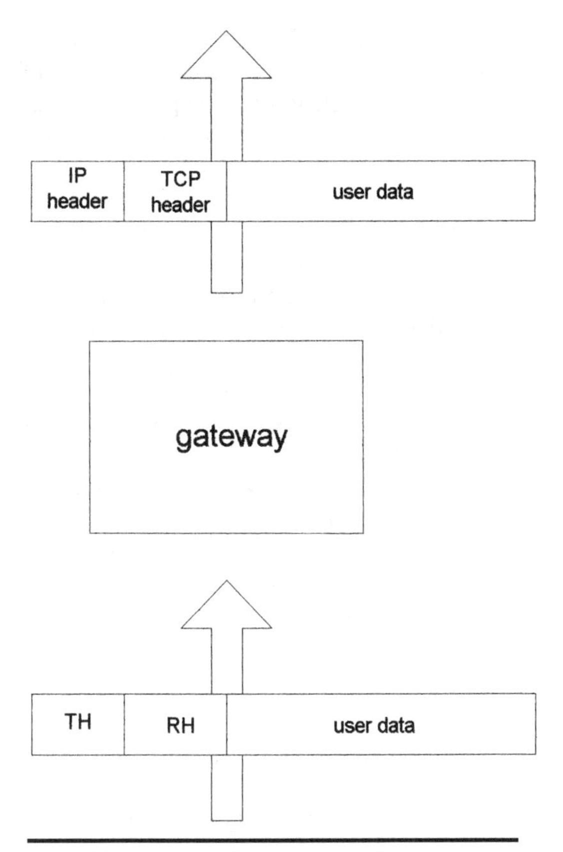

Figure 13.5 Legacy TCP gateway—SNA.

- Aynchronous, low-speed access
- Asynchronous and synchronous point-of-sale devices

All of these interfaces have a lower throughput capability than a token ring or ethernet TCP connection. This is largely a physical limitation of the serial-based communication of the downlink.[4]

Throughput for this type of gateway is quite high because of the isolation of the protocol stacks. In addition, gateway downlink interface is tailored to optimize throughput.[5] Steps are taken to utilize

[4]When a gateway is used for a downlink that is for token ring SNA devices, the throughput capability of the SNA protocol can be seen.

[5]An attempt is made to provide an optimized interface. There are restrictions in the capabilities of the downlink attachment that do not allow an optimized interface.

options within the protocol that reduce the efficiency of communication. For example, the SNA LU2 interface can use a different set of protocol options to optimize the interface. Although this gateway supports several different communication options, it is recommended that only a subset of these be utilized.[6] In this case, use of LUs that specify BB (begin bracket) and EB (end bracket) on every request and response be used. This keeps the LU maximally available without having to send extra frames to maintain proper communication. An example of this is when a frame must be transmitted with only an end bracket indicator (EBI) to end a bracket and allow communication to proceed.

At the same time, the TCP interface has been designed to provide very fast response time and high throughput. Because the TCP session traverses all of the way to the downlink protocol interface, the ability to closely monitor the status of the downlink connection exists.

The TCP end-to-end integrity checks ensure that the data has successfully reached the destination. Because the IEN network operates on datagrams,[7] use of the TCP integrity checks allows communication without requiring additional overhead or delays.

13.3 Security Implications

Security for this type of gateway configuration is a challenge. The most serious reason for this situation is that the security interfaces of the different network protocols are not necessarily compatible, nor are they transportable between the two sessions.

As shown in Fig. 13.6, the security operable on the subarea SNA side is not transportable to that on the TCP side. The subarea SNA interface can use one of several security interfaces. These include interfacing to the security modules within the host system. An example of a product that performs this is Resource Access Control Facility (RACF).

It is also possible to utilize an interface such as the security interface to VTAM and NCP. This interface utilizes encryption to provide security for the communication path.[8]

---

[6]The other options are available for customers that cannot use one of these protocol options.

[7]There are times when end-to-end acknowledgments are included within the Hypercom interface.

[8]All of the SNA protocols can also implement conversation-level security for LU 6.2.

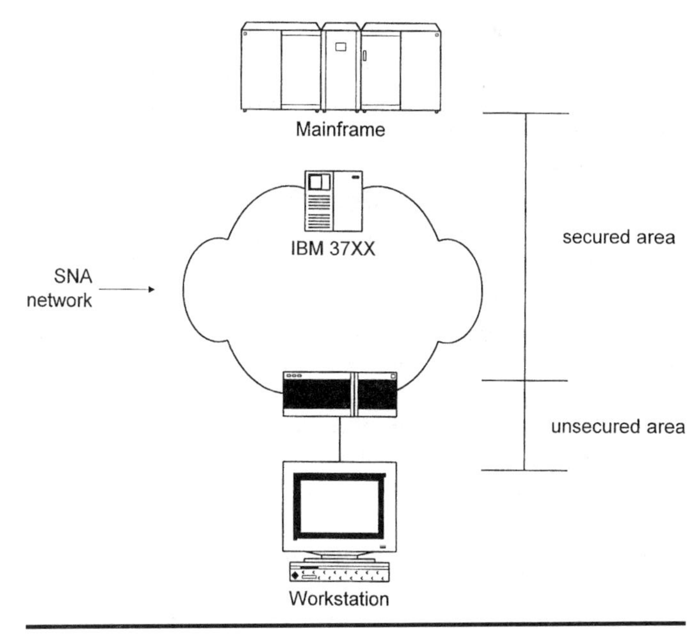

Figure 13.6 Security.

13.3.1 Gateway security issues

A *gateway* is, by definition, a common path between two systems. It is the common characteristic that causes the security issue.

Most gateways provide access to a pool of resources. This pool is normally utilized on a FIFO basis. When a demand arises for access, one of the resources within the pool is used to provide this access. There is no passing of any security information between the two sessions. As a result, no relationship is established between the security of one session and the next session.

13.3.2 Methods of addressing security

A security interface is required that allows either a common access point for both session types or an interface that is accessible from both sessions.

In the first case, a single security system allows security verification for both sessions. By definition, this must be a cross-platform security system that allows accessibility through different interfaces. This is because the system must allow access through an SNA or TCP access point.

13.3.3 Common security interfaces

Several systems fit the profile outlined above. Among these are Kerberos and KryptoKnight. These are both open security systems that are capable of handling requests from a different set of interfaces. Yet, both of these security interfaces provide very strong security that can be transferable between sessions.

For example, IBM offers a security program called NetSp which builds upon the GSS security API. This is a standards-based security that is specified in RFC 1508 and 1509. This program allows security requests from an assortment of access points. Using this program, it is possible to build a secure system that requires only a single logon. This type of feature makes this product a good fit for this purpose.

These products work on the principle of a *trusted server.* The client either communicates directly with the security server or, as is the case of client/server configurations, the authentication server operates as a third party.[9]

With this type of security interface, it is possible to maintain access control and consistency between the pair of sessions. The security of one session can be automatically maintained across the gateway to the paired session.

13.4 Network Management

Some say network management is an oxymoron. Can networks be managed? This is the $64,000 question that network management staff have been attempting to answer for many years. From all appearances, the answer is yes. With this in mind, you must now look at what is being managed and how it is managed.

I will use a liberal interpretation of network management in this section. For this purpose, network management is defined as follows:

> The ability to detect the incorrect operation of the network and the ability to detect degradation in service to network users.

The former objective—to detect errors within the network—is the minimal requirement of network management. This objective is met by the ability to detect an error as it occurs in the network. The errors that should be detectable include, but are not limited to:

---

[9]I will not go into extensive details about how this system operates, since this topic is outside of the scope of this book. The discussion here is not as important as the fact that this type of security interface is possible!

- Datalink error threshold exceeded
- The loss of connectivity to a specific interface
- The loss of a session to an endpoint
- The loss of necessary physical connection to a critical component

When an error is detected, a notification method must be used to ensure that an action can be taken. This notification method can involve notifying a human being who provides additional diagnosis or correction, or notifying an automated facility that can perform the same type of tasks.

13.4.1 Legacy network management

Legacy systems have had a requirement for network management from the beginning. Systems have been created to fulfill this requirement. These systems have largely been successful in the legacy marketplace by providing management for these networks. The requirements for management at this level is very high because of the historically high cost of equipment and the communication resources necessary for networking.

For example, NetView has been a large seller within the IBM mainframe marketplace. This program product has enabled SNA customers throughout the world to obtain both real-time notification of network outages and errors, and has provided a historical database that allows trend analysis to be performed.

NetView uses the CNM (communication network management) exit of VTAM to provide a primary program operator (PPO) to allow both solicited and unsolicited messages to be displayed on a NetView operator screen that has the correct authorization. Use of the PPO interface allows a NetView operator to obtain all of the management information that is available to an operator viewing the operating system console.

The outcome of this is that the NetView operator becomes an extension of the normal operation of the host system. All abilities of the operating system console are extended to the authorized NetView operator.

13.4.1.1 Legacy operation. Operation of the network is moved from the operating system console to the NetView terminal on someone's desk. This provides better accessibility to the information and allows for quicker intervention and improved availability of the network.

NetView also has the capability to perform automated tasks. These tasks can be caused by any of the following:

- Messages
- Change in status of a resource
- Asynchronous operation relative to a specific event in the network

In addition, Netview can be further extended in operation through the use of user exits, command processors, and user tasks. Each of these elements provides different types of capabilities; together, they extend Netview's ability to manage networks by adding new capabilities.

Many users have made use of these interfaces to provide such enhancements as the automated reacquisition of network resources, creation of an enhanced message of a network event, or providing a programmatic interface for such applications as trouble ticketing.

Taken together, this network management platform and enhancements provides a unified environment providing a strong management approach with the possibility of keeping an eye on the network while allowing the management platform to provide many enhanced functions.

13.4.2 LAN network management

There are different types of network managers for LAN networks. They may basically be broken down into two types:

- Proprietary systems
- Standards-based systems

13.4.2.1 Proprietary management systems. Proprietary systems were created as a private method of obtaining network management information. These systems operate by allowing information to be obtained on errors that are seen by the LAN interface. These error events are also logged into a statistical database to allow trend reports to be created.

In addition, very specific errors are identified within the LAN interface. These errors or events are used as diagnostic tools for the interface itself. As such, these are not external but internal events that may result in external consequences.

These systems are normally fast and are able to extract a great deal of information from the network and the interface. The speed of this extraction method it is tailored to the specific requirements of the task. These are normally not extensive, generalized information extraction interfaces.

These systems also tend to provide a more efficient method of obtaining information on the network management system itself. By being designed specifically for the management system, the proprietary system may be optimized for operation with the management system.

The proprietary system will also not be handicapped by the overhead that can be associated with standards-based solutions.[10] Although the interface is not standardized and cannot necessarily interoperate with other systems, the interface is usually fast.

13.4.2.2 Standards-based management systems. There are two major standards for management systems. These are:

- SNMP systems
- CMIP systems

In this section I will discuss systems based on Simple Network Management Protocol (SNMP).[11] SNMP is an IETF standard method of obtaining management information. SNMP is based on a requester/server model. Each component in the network can provide management information to an SNMP management system.

Information is architected into formats known as management information bases (MIBs). There are standard and private MIBs. The standard MIB contains information that has been architected and chosen by the standards bodies. A private MIB contains information that is specific to an interface. This MIB provides information that is special to this interface and/or vendor.

The MIB is formatted into an encoded form that allows for self-defining variables. This form is called ASN.1 encoded data. By using standard but expandable methods of encoding information, both standard information and information specific to a product can be transported and understood by the endpoints of the management interface.

SNMP management systems and agents have been exploding in availability over the last few years. This has been brought about as a result of several changes. These include:

---

[10]These types of standards are based on the idea of a common denominator. This denominator may not provide the cleanest methodology for operation.

[11]The other protocol, CMIP, is an open systems interconnection (OSI)-based management protocol that allows a similar interface to SNMP.

- The creation of more multivendor networks
- The increased use of SNMP support in diverse network products
- An attempt to create a common interface for management without having to support multiple management stacks

13.4.4 Bringing legacy and LAN management together

As has often been touted, it would appear that the simplest way to create a common management platform is to pick a protocol that both sides can support. From outward appearances, this would make sense. If both can, for example, use SNMP, you could pick SNMP as your base.

This could work if the control that is provided by the SNMP system were equivalent to that of the legacy system. If you could provide the same information at the same speed with the same responsiveness, the world would be perfect! Unfortunately, this ideal does not match up with reality.

SNMP management in the legacy environment does not currently fulfill the ideal of "management for the masses." SNMP can be supported through the use of external interfaces that allow the nonnative data to enter. Among the methods for support is through MultiSystem Manager for NetView, NetView/6000, and some of the TCP products that operate on the host, such as TCP/IP for MVS. These products allow SNMP data to be translated into a form that NetView on the host can support.

The overhead of SNMP, though better in the current architecture than the original, can still be quite large. This overhead is caused by the polling nature of SNMP. When the number of nodes being managed is small, extensive bandwidth is not required. When this model is expanded and imposed on a large network of, say, 10,000 nodes, a polling cycle of even 5 minutes requires sending 33 messages per second for management alone. There are ways of reducing this impact, but I think you can understand the impact of having centralized management and control.

13.4.4.1 Divide and conquer. The simplest method of "integrating" management techniques is to provide a human integration by providing both of the management interfaces without using any cross-connection. In this scenario, you would provide a network manager with two screens, one for legacy and one for LAN. Some users might feel that this allows each platform to provide support for what it does best. Although no integrated view of the network can be displayed, this is of little consequence as the points of interaction are well known and can be quickly overlaid by a human operator.

Although this is an extreme solution, it may be the best alternative for some users. These cases include, but are not limited to:

- When there is sufficient personnel to provide full support for both solutions
- When the networks are not truly interconnected
- When there is insufficient personnel to provide a clean migration to a single system
- When a feature is required that cannot be integrated or duplicated by the other interface

Thus, although this would not appear to be a good solution,[12] it may be an acceptable one for some users.

Moving up the scale of integration we find the solution of only integrating some functions. This scenario provides for the continued separation of the management of the networks into isolated components. The difference is that some of the management information is integrated into a single point. This limited integration may be for only a few alerts or counters. In either case, integration has moved forward from the previous solution.

The integration could be as simple as a shared file into which certain data is written. In this case, it would be helpful if the data were written to the file in the format the other system is expecting. This would allow the partner system to read the information as if the data were written by the original system. Trend reports, integrated views of the network, or other types of integration could be utilized within this configuration.

This is a viable method of migrating into a fully integrated environment. This method provides for some interaction, but reduces the use of creating a fully integrated solution.

Since this solution is flexible enough to allow differing levels of integration, you can start with a very limited level of integration and move through each information piece until you are ready for a fully integrated solution.

We have now gone full circle and are looking at a fully integrated network management system. The difference is that here we will address some of the issues that were previously brought up.

---

[12]It definitely is not an integrated one!

The first task is to determine what underlying management architecture to use. Since there are SNMP interfaces for host-based management systems, whereas the reverse is not often available, we will use an SNMP architecture as the basis for our consolidated management system.

As has already been explained, use of standards-based management for legacy systems can make management more difficult because it is not optimized for this use. Instead it is based on the least common denominator. But there are options that correct this shortcoming. These include using the SNMP platform for some portion of the total network management task. For example, you could use an SNMP product like NetView/6000 to deliver messages of status change for various systems to a single screen. In this case, the messages are created by the legacy management system, but management responses are sent through the proprietary interface.

Another alternative is to obtain all of your asynchronous alerts[13] through the SNMP system. Statistical information would be solicited through the proprietary system to reduce the management overhead through the communication interfaces. An alternative method would be to greatly reduce the statistical poll time. For example, it would be done once per hour.

13.4.5 Additional features

The integration of network management systems requires more than just the use of a different management methodology. Additional steps should be taken to ensure a smooth integration path. These steps are not required, but are strongly recommended to protect the user's investment of time and effort in the existing management system.

13.4.5.1 Integrated configuration database. All management systems include a database covering the configuration of the network that is being managed. This database includes such information as:

- The physical location of the resource
- The name of the resource
- Special capabilities
- The interrelation to other components

Together, these pieces of information define how the specific resource is integrated into the network. Without this information, it

[13]These alerts would come through the trap interface of SNMP.

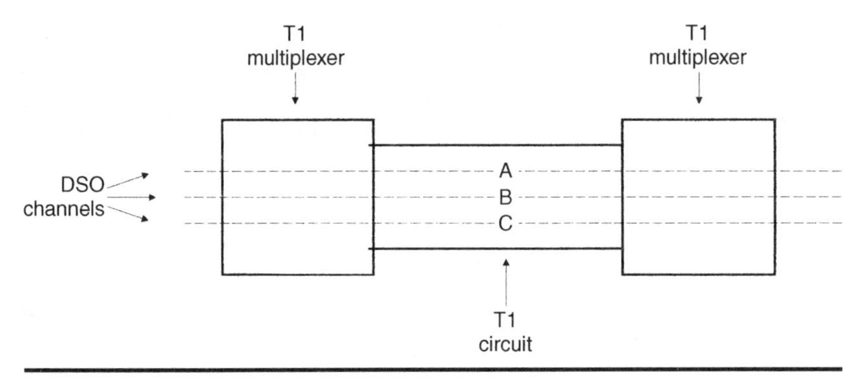

Figure 13.7 Physical and logical network management.

is very difficult to determine the effects of an alarm and how different alarms may be related to each other.

Figure 13.7 shows how logical and physical networks can be related. In this figure, you can see that the simultaneous alarms for lines A, B, and C could imply that the T1 multiplexer or T1 line has failed. If an alarm is received due to the failure of the multiplexer, you can be assured that this failure is at least a major portion of the cause for the failure. If the T1 multiplexer shows an operable state, then attention can be shifted to the T1 line. These concurrent events do not mean that the failures are actually more serious than they appear, but the probability starts to decline and point toward a "higher level" failure.

Without an integrated configuration database, it would be difficult to make these types of deductions because the information would not be readily available. Instead, you would be attempting to track down three different problems instead of the single problem of the failure of the T1 multiplexer.

13.4.5.2 Integrated problem reporting. Problems occurring in the network must be both trended and monitored. In large networks, the ability to produce trend reports can be severely limited. Without the ability to produce trend analysis, it can be difficult to determine the health of the populous.

Problems from different management domains must be integrated into a single reporting structure that facilitates analysis and reporting. The single repository must report the problem into a simple, single format. The details of that format are less important than just that it exists; data manipulation can always be done.

The use of a true database for the problem reports would have certain advantages for data extraction, though providing sufficient input speed could be a problem. If necessary, the data can be input to a database in a batch process. This reduces the overhead on the data writes while also providing for better analysis from the use of a database for extraction. Once data is contained in a database, data extraction can be done much more easily, more quickly, and with more variation.

13.4.6 Network management summary

As you can see, consolidation can be achieved, but there are several obstacles that must be navigated before true integration can result. Among these are:

- The integration task must be planned carefully and must provide interfaces so that data can be shared or directed toward a common database.
- An interface that provides a common integration platform must be defined and used by all management interfaces.
- Because of possible overhead associated with running a standards-based management, a proprietary interface may be used instead.

13.5 Summary

The integration of these network architectures, although highly desirable, does raise areas where concern is both expected and appropriate. These areas arise because of the fact that events occur on an integrated system that are normally not existent. For example, devices may exist that are not natively supported in the case of unexpected events.

The area is that of the shared access path. By integrating different protocols onto a common access path, it is possible to save money, by reducing the physical paths that must be purchased. However, sharing the access requires planning and proper management to ensure that it proceeds smoothly and that the protocols do not reduce the effective throughput of the interfaces.

It must also be understood that the creation of a shared access path does not equate to the creation of an integrated network. Although different protocols share a communication path, in all of the multiplexing methods, this is of little gain toward integration in the old hardware multiplexer. All that is gained is the number of physical paths that must be supported. The protocols are still isolated as before—they just utilize a common path.

At the same time, there are several benefits to creating a shared access path. Among the largest, and the first one I discussed, is monetary. Communication paths are not cheap. Although the costs have come down, they can still amount to millions of dollars for the largest companies. At these levels, even a small reduction can amount to spending "real money."

Another benefit, and one that is not often discussed, is an outcome of investigating one's network. Doing this results in gathering information on the network, its users, and its usage. Often companies are unaware of how their networks are used. It is common for a simple analysis leading toward an integration task to result in the detection of new uses for the corporate network.

An integration effort can lead to an integration of users. This occurs because information is gleaned about the network users that was not evident before. For example, users may find out that another group has information that is valuable to others. In this case, companies gain through the simple cross-awareness that can result from an integration effort. Often one group learns that others exist and that cooperation can be mutually beneficial.

Gateways are often used when true integration of protocols is an outcome of an integration effort. In this case, a gateway is constructed between two protocols that normally cannot communicate. This type of protocol gateway is very effective as a method of creating the illusion of a contiguous session that crosses a network protocol boundary. The users of the gateway are unaware that there are actually two sessions.

A problem that arises from this type of gateway is the impact on security. There are two issues in this area that cause concern. The first issue is to ensure that the status of one session is reflected in that of the other session. Thus, if one session becomes unavailable, the other session should experience a similar loss of connectivity.

This is not a difficult task to accomplish, but many factors make it both expensive and difficult to perform. The first factor is that the termination and reestablishment of a session is expensive in terms of both network and system resources. As a result, it is normal (and appropriate) to delay the deactivation of a session by the currently active session in order to provide an opportunity for the first session to become active again. This direction relies upon the fact that attempts are made to ensure that session resources stay active. As a result, it is the norm to attempt to keep sessions active and, thus, it is appropriate to wait for the reactivation of the session.

In some configurations, it is difficult to gain knowledge if the first session has failed. An example is a TCP session that is not carrying any traffic. Because it is normal for keep-alive timers to be set to 30 minutes, it is possible for a session to fail without users being aware of this for an extended period of time. In these cases, the configuration can actually go through several iterations of active/inactive states without anyone knowing about it.

The second issue regards gateways, in the case where the security profile of one session does not automatically transfer to the other session. As a result, it is possible that inconsistent security controls exist because of a possible inability to determine security on one of the sessions.

If the gateway provides a pool of resources to be used by downlink sessions for a configuration that is very common, the security characteristics of individual downlink sessions will not be honored or consistent. A possible solution is to use a consistent security interface, such as one of the standards-based facilities like Kerberos or KryptoKnight.

Another major concern involved in this type of integration is that of network management. This is an area of conflicting issues that must be addressed.

Network management can be done with either proprietary or standards-based tools. The SNA entities have historically been managed by legacy, proprietary tools, such as NetView.

The legacy, proprietary management tools are tailored to provide information that can optimally be obtained by the resources within the network. In addition, these resources have been tuned to provide the information in a specific way.

Although these are "proprietary" methods, it is quite common for others to build interfaces, if the market is large enough, as is the case of the large mainframe environments in which NetView operates.[14]

SNA management can be provided quickly and effectively by using NetView. The flows involved have been optimized for use in this environment. The ability to obtain both synchronous and asynchronous events gives network control operators the capability to react quickly to events within the network.

LAN network management has been historically done through the use of standards-based management protocols. In most cases, the protocol used is SNMP. SNMP is an IETF management protocol that allows a diverse set of resources to be managed and inquired.

---

[14]Although NetView is not a truly proprietary interface, the interface is so complex that it is often viewed as such. In addition, it is not based on one of the standards-based architectures.

SNMP uses a client/server model for operation. In addition, it expects the server to poll clients for their status. This results in an efficient use of network resources when the number of resources being managed is low. However, when the managed resource number rises, the overhead of information polling can become considerable.

When applied to legacy resources, SNMP requires a higher level of network overhead and resource cycles. Because it was designed to provide management from a diverse set of resources, it is not necessarily optimized for the types of network hosts. The overhead of polling 10,000 resources—a number that is not uncommon—can exceed 33 messages per second. This is a large number of requests; in the case of network management, there is no direct benefit from SNMP data polls. Instead, this data prevents useful data from traversing the network.

At the same time, there are several driving factors behind the move to achieve an integrated management platform. The prime factor is that such a platform would aid the network management personnel in obtaining and using data from the network. By using an integrated system, it is possible to obtain an integrated view of the network and to view the network from one interface. There is no reason to have more than one screen depicting the network and its resources.

Several methods of providing this integration were presented. Among the methods discussed is incremental integration. This method allows different management platforms to manage different resources, but allows for the existence of some integrated message and command interfaces.

The integration of the management platforms is facilitated by the methods discussed in this chapter. Together, most of the concerns of an integrated network have been shown to be surmountable. The concerns, although quite real and important, can be overcome by reasonable methods that allow costs to be monitored and controlled.

Glossary

AIW APPN Implementors Workshop.

ANR Automatic Network Routing. A method of routing provided by HPR nodes.

ARB Adaptive rate-based congestion control. ARB provides preventative congestion control by providing control on the speed data enters the network.

ARP Address Resolution Protocol.

ARPA Advanced Research Projects Agency.

ARPANET A network that was created by the U.S. Department of Defense in the early 1970s. The precursor of the current Internet.

ASM Address space manager.

BECN Backware explicit congestion notification. An indicator in the FR network to notify downstream devices of network congestion.

BIND An SNA command that establishes a session between two LUs.

Boundary node An APPN node that allows connection to a non-native network

CDRM Cross-Domain Resource Manager.

CDRSC Cross-Domain Resource.

Chaining Session-level packetization scheme in subarea SNA and APPN.

CICS Customer Information Control System. CICS is a transaction monitor that operates on top of VTAM. Users can create transaction programs that interact with either remote terminals or other computing systems, such as other CICS systems.

CIR Committed information rate. An agreed-upon rate of data transport for frame relay networks.

Class of service table Table that specifies the set of output queues that a session will use. This table also maps to the virtual routes in subarea SNA to specify such session characteristics as the use of encryption.

CLIST Command list. This is an interpreted application interface. CLISTs are used to perform automated tasks within an application, such as a network management program, like NetView.

CMIP Common management information protocol. A network management protocol that is part of the OSI suite of protocols.

CNM Communication network management. VTAM provides a CNM interface that allows network management data to be solicited from downstream devices. The CNM interface also support the asynchronous receipt of management data.

Composite node A composite node is formed by a combination of VTAM and NCP working in cooperation to provide the appearance of an APPN node to others in that network.

Control point (CP) A controlling resource within an APPN or LEN network.

COS Class of service. This is a subarea SNA and APPN configuration item that defines the type of services required of a session.

CSMA/CD Carrier Sense Multiple Access/Collision Detection.

DAF Destination address field. The address of the destination of a session.

DE bit Discard eligibility bit. A bit that designates that a frame may be discarded by a frame relay network.

DLC Data Link Control.

DLCI Data link connection identifier. A frame relay virtual circuit identifier.

DLSw Datalink switching. A standard method of allowing the transport of SNA and NetBIOS across an IP network. RFC 1434 and 1795 define this standard.

DLUr/s Dependent LU Requestor/Server. An extension to base APPN to allow support for dependent LU access and utilization across an APPN network.

Domain A section of a network. In relationship to subarea SNA, it defines the span of ownership for network components.

End node (EN) A full APPN node that does not provide route calculation features.

ER Explicit route. The physical path between two adjacent subarea nodes. Explicit routes define the path upon which a VR is defined.

ESCON Enterprise system connection architecture. A fiber channel to IBM systems.

FECN Forward explicit congestion notification. An indication in the FR network to notify upstream devices of network congestion.

FID Format Identifier.

FIFO First in, first out. A queuing method.

FR Frame relay.

Fully qualified procedure correlation identifier (FQPCID) Network unique identifier that identifies a session. This identifier is network qualified.

Gateway Software interface between two network architectures.

GDS Generalized datastream. A structured data field. There are two formats of GDS. These are key length (KL) and length key (LT).

HDLC High-level datalink control. A standard DLC that utilizes a bit-oriented encoding of data.

High-performance routing (HPR) An extension to the APPN architecture. This architecture, which is still being designed, will allow session paths to be dynamically modified according to network conditions and availability.

IBM International Business Machines. An international computer vendor that designed subarea SNA and its follow-on architectures, APPN and HPR.

ICMP Internet Control Message Protocol.

IEN Integrated Enterprise Networking. A group of communication equipment offered by Hypercom Network Systems. These products range from small access devices to large, central processing equipment.

IESG Internet Engineering Steering Group. The IESG coordinates the IETF working groups.

IETF Internet Engineering Task Force. A group of the IAB (Internet Architecture Board). The IETF leads working groups in immediate requirements of the Internet.

IMS Information Management System. IMS was originally a database interface for batch programs. It evolved into a transaction monitor, like CICS.

Independent LU An independent LU does not require the services of an SSCP to establish an LU-LU session. The minimum base to support this is a PU 2.1.

Internet A loose collection of networks that comprise systems throughout the world. It is the outgrowth of the defense industry ARPANET.

IP Internet Protocol. The network layer used for TCP/IP.

IRTF Internet Research Task Force. A group of the IAB (Internet Architecture Board). The IRTF establishes working groups for long-term research.

ISDN Integrated services digital network.

ISO International Standards Organization.

ISR Intermediary Session Routing. An APPN component that creates a routing component along the path between two APPN nodes.

KL A format of GDS variable. In this format, the key field precedes the length field. The length does not include the key field.

LAN Local area network.

LFSID Local form session identifier.

Limited resource path A switched facility that is only meant to provide connectivity for short durations. When the usage count drops to zero, this type of facility is meant to be terminated.

Link header This is the header for a specific DLC. In the case of SDLC, this consists of the following parts: the X′7e′ flag, pole address, and link control sequence number field.

LMI Local management interface. An interface to a frame relay network that allows information at the subscriber network interface (SNI) to communicate, bidirectionally, as to the status of the partner.

Local form session identifier (LFSID) A local identifier created by the address space manager for each session from an LU. *See* Fully qualified procedure correlation identifier (FQPCID).

LOCATE GDS GDS variable used to communicate between the originator of a session and the destination. The LOCATE GDS variable contains additional vectors.

Low-entry networking (LEN) An implementation of APPN that does not include either the directory or dynamic routing logic of end nodes and network nodes.

LT A format of GDS variable. In this format, the length field precedes the key field. The length includes the key field.

LU 6.2 session A peer-to-peer session type. Used between the control points (CPs) of APPN nodes.

MAC Media access control. For LAN connections, this defines the physical addressing across a LAN connection.

MIB Management information base. This is an ASN.1 encoded datastream for network management in SNMP.

MLTG Multilink transmission group. An MLTG is a logical grouping of multiple physical circuits.

MPTN Multiprotocol transport networking. This is an architecture developed by IBM to allow different protocols to share a common transport.

NCE Network connection endpoint. A termination point for an RTP connection.

NCP Network control program. NCP operates within the IBM front-end communications processor, such as the 3745.

NETID Network identifier. The NETID establishes the name space for locating a resource.

NetView A network management interface that operates on all mainframe operating systems. Versions of NetView are also available for a diverse set of operating environments, including UNIX and OS/2.

Network node (NN) A full APPN node that provides all base-level services of APPN.

NHDR Network layer header. Provides addressing for the packet as it traverses the HPR network.

NLPID Network layer protocol identifier. This identifier is administered by ISO and CCITT and provides a standard method of dynamically determining what type of data is being carried within a frame or virtual circuit.

NMVT Network management vector transport. This network management command allows formatted information to be transmitted in both directions (to and from a network management interface).

NOF Network operator facility. One of the fundamental components of an APPN node.

NRF Network Routing Facility. A program product that allows secondary LUs to establish a session.

OAF Origin address field. The address of the origin of a session.

ODAI OAF/DAF assignor indicator. This is a binary field that defines which session partner chose the address for a session. This is used by APPN nodes.

OSI Open Standards Institute. International standards organization.

OSPF Open shortest path first. A routing algorithm that allows the routing domain to be subdivided to allow routing servers between domains.

OUI Organizationally unique identifier. A standard identifier. This is used in several interfaces, including token ring and frame relay.

PPP Point-to-Point Protocol. A method of interfacing multiple protocols across a serial circuit. PAP defines the use of a protocol identifier to determine which protocol is being transported. PPP supports the use of either asynchronous or synchronous serial circuits.

PC Path control. PC is the APPN component that allows for the initiation and termination of network paths.

Peer-to-peer Connection at a peer level, as opposed to a hierarchical configuration.

PID Protocol identifier. A standard method of defining the protocol being carried across an interface.

Priority routing An option set supported by some TCP/IP routers. It allows certain traffic to be given higher priority in routing.

PSID Product set identifier.

Quiescing The process of bringing a system to a stable state. This normally applies to bringing a system to a closed state.

RARP Reverse address resolution protocol.

RFC Request for comment. A formal descriptive of a protocol. Made publicly available for others to comment on and add to.

RH Request/response header. A 3-byte field that immediately following the TH. It allows remote facilities to agree on some interface with finite-state machines.

RIG Related interest group. A group within the APPN Implementors Workshop. One example is the RIG developing DLSw.

RIP Routing Interface Protocol. RIP is used in IP to determine the path to a destination.

RJE Remote job entry. This network architecture allows remote devices to enter batch data. Examples include 3780 BSC and 3770 SNA devices.

RNR Receiver not ready. A command used to provide the local acknowledgment number to the DLC partner and to signal that transmission must be stopped.

RR Receiver ready. A command used to provide the local acknowledgment number to the DLC partner and to signal that transmission can be continued from the acknowledgment point. SDLC and HDLC use this link command.

RSCS Remote spool communication subsystem. RSCS is a component of the IBM VM (virtual machine) operating system. It provides the communication vehicle between virtual machines.

RTP Rapid Transport Protocol. A full-duplex, connection-oriented protocol that is designed to support high-speed, low-error-rate communication paths.

RU Request unit. The user data that is transported across an SNA network.

SAP Service access point. SAPs allow for the type of data being transported to be easily identified.

SDDLU Self Defining Dependent LU. By passing the PSID (product set identifier), VTAM is able to dynamically create a definition for a dependent LU.

SDLC Synchronous Data Link Control.

Segmenting Transport-level packetization scheme in subarea SNA and APPN.

SIDH Session identifier high. The high 8 bits of the LFSID.

SIDL Session identifier low. The low 8 bits of the LFSID.

SLIP Serial Line Interface Protocol. A method of interfacing a protocol across a serial circuit. SLIP uses a few escape sequences to allow the transmission of binary data. When using SLIP, each end of the circuit must agree upon what protocol is being supported.

SNA System Network Architecture. Network architecture designed by IBM. Often referred to as subarea SNA. This architecture was expanded as APPN and further expanded as HPR.

SNA/MS SNA Management Services. An architecture that allows a management interface that exceeds the capabilities of CNM.

SNAP Subnetwork access protocol. A header type for IP data.

SNI SNA network interconnect. A method of interconnecting subarea SNA networks.

SNMP Simple network management protocol. A standard method of providing network management. Management data is obtained through the use of management information base (MIB) encoded data.

SNRM Set normal response mode. This is a link-level activation command used in SDLC.

SSCP System services control point. An SSCP defines a PU type 5 node. VTAM is one of the few SSCPs.

SSE System Services Extensions.

STUN SNA tunneling. A feature offered by cisco systems to pass SNA data across an IP network.

TCP Transmission Control Protocol. One of the transport protocols used in TCP/IP. TCP provides a *session* between the endpoints.

telnet The telnet application allows a client system to log onto another system as a terminal.

TG Transmission group.

TH Transmission header. The TH follows the DLC link header. SNA defines three types of TH. The first is the FID2, which is 6 bytes long. This type is used for connection to PU2.0 and PU2.1 nodes. The second is the FID4, which is 26 bytes long and is used between subarea VTAM and NCP. The third type is the FIDF, which is only used between HPR nodes.

THDR RTP transport header. Used by the RTP endpoints to provide correct processing of the packet.

tn3270 This program uses the binary mode and end-of-record codes to allow 3270 datastreams to be sent to a terminal type that is not normally utilized, such as VT100.

TRS Topology and Routing Services. One of the key components within an APPN node. TRS calculates the route to a destination.

TSO Time Sharing Option. Adjunct to the IBM MVS operating system that allows system support, such as text editing and simple file management.

UA Unnumbered acknowledgment. This is a link-level acknowledgment to a link activation command. An UA is used to respond for an SNRM, for SDLC, or an SABM, for HCLC.

UDP User Datagram Protocol. One of the transport protocols used in TCP/IP. UDP provides a *connectionless* connection between the endpoints.

Virtual circuit A method of multiplexing datastreams. This technique is used by such interfaces as X.25 and frame relay to simulate point-to-point lines.

VC Virtual circuit.

VR Virtual route. The logical path between two SNA subarea nodes. A VR defines the end-to-end path. A VR flows across explicit routes. *See* Explicit route.

VTAM Virtual Telecommunications Access Method. VTAM is the centerpiece of a subarea SNA network. It is an access method that operates under all IBM mainframe compatible operating systems.

WAN Wide area network.

XID Exchange Identification. Frame to identify a node to the network.

XID3 A key component of the PU 2.1. The XID3 allows for the transportation and negotiation of station parameters.

Index

ABOUT THE AUTHOR

David G. Matusow is a network professional with hands-on experience designing and installing subarea SNA, APPN, and TCP/IP networks. This work includes the development of network software for clients throughout the world. His work includes design work for IBM that resulted in the release of VTAM Version 4, the first APPN-capable version.

He is currently the Chief SNA/APPN Architect for Hypercom Network Systems. In this capacity, he is responsible for setting the direction for the company for this segment of the industry. He is also responsible for the development and support of the SNA and interconnection interfaces for the Integrated Enterprise Networking (IEN) product line.

Mr. Matusow has been extensively published in the communications industry, as well as developing and providing seminars for customers throughout the world.